Plant Growth Curves

The Functional Approach to Plant Growth Analysis

'In general, curve fitting should be indulged in only when there are clear-cut objectives, and when the practitioner is aware of the pitfalls'.

(Williams, 1975)

'. . . *omnes observationes nostrae propter instrumentorum sensuumque imperfectionem non sint nisi approximationes ad veritatem* . . .' ['all our observations, because of the imperfection of instruments and of the senses, are nothing more than approximations to the truth'].

(Gauss, 1809)

Plant Growth Curves
The Functional Approach to Plant Growth Analysis

Roderick Hunt
B.Sc., Ph.D., M.I.Biol.

Independent Research Worker, Natural Environment Research Council Unit of Comparative Plant Ecology, Honorary Lecturer in Botany, University of Sheffield

University Park Press
Baltimore

First published 1982
by Edward Arnold (Publishers) Limited, London
First Published in the U.S.A. in 1982 by
University Park Press
233 East Redwood Street
Baltimore, Maryland 21202

British Library Cataloguing in Publication Data

Hunt, Roderick, *1945–*
Plant growth curves.
1. Growth (Plants)
I. Title
581.3'1 QK731

LCCN 82-50119
ISBN 0-8391-1757-4

Printed and Bound in Great Britain

Preface

Plant growth curves in the special, but far from idiosyncratic, sense of this book are progressions of plant size against time. They have long occupied a vital place in plant science where studies on the performance of whole plants have often acted as the centre from which work at other levels of organization has radiated. The notion of change in size, too, with its fascination for the human mind and its indispensible involvement in biological sciences, has attracted statisticians from the earliest days of their subject and now occupies, through the agency of the fitted curve, a special place in the quantitative analysis of plant growth.

This book reviews the theory, practice and applications of fitted plant growth curves, a project I undertook both with excitement and with sadness; excitement, because this is the first time that a reasonably comprehensive coverage of this gratifyingly concise topic has been attempted, and sadness because, in view of the explosive development of the literature, it will in all probability also be the last, subsequent work necessarily involving some further subdivision. The time for viewing the subject as an entity is ripe, but this will soon pass as its twigs become branches, and so on.

Such matters as the book's background, anticipated readership and interrelationships with existing publications I cover in a separate introduction.

The work's existence owes much to many and my special thanks are due to Dr D. R. Causton for his friendly and critical co-operation at all stages of the project; to Dr G.C. Evans, who has on innumerable occasions given generously of his matchless experience; and to I. T. Parsons, a friend from school days and a valued collaborator over the last ten years.

My interest in plant growth analysis was first awakened at Sheffield by Professor A. R. Clapham. Since then, many friends and colleagues at the Universities of Bristol and Sheffield have shared with me their problems, ideas and successes in this and related fields: Dr J. F. Hope-Simpson, Dr E. I. Newman, C. R. Baines, the late Dr P. S. Lloyd, Dr J. Grace, the late Dr R. Law, Professor A. J. Willis, Dr I. H. Rorison, Dr J. P. Grime, Dr P. L. Gupta, Dr A. J. M. Baker, Dr H. Gretton, Dr S. B. Furness, Mrs. S. A. Fathy and Dr S. P. McGrath. Friends, collaborators and correspondents from many parts of the world have also brought me into contact with many diverse topics related to the study of plant growth, their numbers swollen by my good fortune in acting as Liaison Officer to the Natural Environment Research Council Unit of Comparative Plant Ecology at Sheffield. Chief among these are Professor F. A. Bazzaz, Dr M. J. Chadwick, Dr P. D. Crittenden, Dr J. G. P. Dirven, the late Dr A. P. Hughes, Dr R. G. Hurd, Dr B. C. Jarvis, H. T. Khong, Dr J. Květ Dr A. S. Mahmoud, Dr J. H. Mook, P. Sivan, Dr D. P. Stribley, Dr. A. Troughton and Dr F. I. Woodward. To these and all others not mentioned, my grateful thanks.

Over the years I have been helped by a series of willing and able assistants: R. J. Allen, Miss C. R. V. Rathey, Mrs H. J. Hucklesby and A. M. Neal while, almost throughout, Mrs A. M. N. Ruttle has provided incomparable secretarial support.

The permissions of the authors and publishers of the illustrative examples are gratefully acknowledged and I am indebted to the cited publication by S. Makridakis which prompted me to search the *Theoria motus* for the Gauss quotation which plays Charybdis to the Scylla of R. F. Williams opposite my title page. Dr D. R. Causton improved the whole text by his lengthy series of reader's comments and Dr G. C. Evans responded in characteristically far-sighted style upon receipt of parts of Chapter 8. Errors and omissions remain my own and I will gladly receive due notice of these from readers, for which my thanks in advance.

Sheffield, 1982 R. H.

Contents

Introduction

This book is heretical for two reasons. Firstly, although it is written by a biologist for the attention of other biologists, its subject matter is strongly mathematical and statistical. Secondly, it advocates an expansion in the use of the empirical model which simply 're-describes' observational data, thus – not rejecting but – forcing back into coexistence the more fashionable mechanistic model with its offers of insights into how living systems work. In making no apology for this standpoint, I suggest only that it is justified because it is what the subject presently needs. Let me elaborate.

The British school of plant growth analysis, which had its origins in the work of F. G. Gregory, V. H. Blackman, G. E. Briggs and co-workers, in their turn drawing some inspiration from nineteenth-century German work, has now been in existence for some sixty years. The method of this school, which amounts to nothing more than a special way of looking at the growth of whole plants, has to varying degrees found a place in the publication of thousands if not tens of thousands of scientific studies. The appeal of the method has always been that useful information may be obtained in amounts out of all proportion to the original outlay in equipment and experimental effort. Though no longer wholly in the van of plant physiological research, the method retains a permanent usefulness in two areas. Firstly, where the frontiers of knowledge about the growth of plants lie closest, as with unexplored experimental subjects or unusual environments and, secondly, where the unique advantages of the method are needed either *per se*, or as preliminaries or adjuncts to other studies of plant growth, be they ecological, genetical, physiological, biochemical or, particularly, agricultural, where the method has quite possibly received its greatest single use.

In the present book I set out the relevance to this field of the so-called 'computer revolution'. There is (almost) no branch of biology into which the advent of high-speed digital computers has not made substantial inroads. As I write, the 'lower end' of this vast field is becoming ever more extensive, with ever more computing power in pocket and desk calculators becoming, in real terms, ever more cheaply available. The purpose of my text is to point out the opportunities that this state of affairs provides, and to review the possibilities for progress in what Causton (1967) has called the 'functional' as opposed to the 'classical' approach to the subject.

The reader I have in mind is simply the experimenter who grows plants and makes sequential measurements on them. He is the student at college or university who is undertaking a project which involves some form of quantitative comparison between experimental subjects or treatments, or the research student or established worker, maybe not principally in this field at all, who has a similar requirement. The standard of mathematical background needed will in algebra and geometry be less than O-level, and in the differential and integral calculus, only elementary. In statistics, familiarity with one of the cited statistical texts for biologists will suffice, with more advanced reading only being necessary – and then not always – in the field of regression analysis.

I have no doubt that to present this text from the point of view of the working biologist may attract criticism, but I can shelter impudently behind the brave stand made by works such as Kenneth Mather's *Statistical Analysis in Biology* (1964) and J. Maynard Smith's *Mathematical Ideas in Biology* (1968). The organization of this text serves the needs of plant growth analysis only; statistical and mathematical ideas are introduced in illogical sequence and are then often incompletely specified. These cupboards are raided greedily in the interests of the experimenter studying plant growth. Nonetheless, it is my hope that, provided the reader also pursues such first-rate statistical texts as the aforementioned Mather (1964) or Norman T. J. Bailey's *Statistical Methods in Biology* (1964) or Geoffrey M. Clarke's *Statistics and Experimental Design* (1980), and David R. Causton's *A Biologist's Mathematics* (1977) (which is uniquely valuable because of its particular emphasis on the process of gröwth), he will not go far astray. J. N. R. Jeffers's first two 'Statistical Checklists' (on design, 1978, and on sampling, 1979) will help, and if all else fails there is always Moran's (1974) bibliography of statistical bibliographies to enable lost ways to be re-found.

None of this is to say that specialist advice, particularly from statisticians, is unwelcome. This is a truth so obvious that it can be overlooked, for statisticians as *partners* in biological research are still comparatively rare (Finney, 1978). When P. J. Radford in 1967 wrote about 'Growth analysis formulae – their use and abuse' the statistical content of his paper amounted to no more than crumbs from the table and yet, as a source of information to biologists, this paper has been seized upon eagerly: to June 1981 it has received a total of 152 citations (*Science Citation Index*). Radford's publication is obviously a 'Desert Island Reprint' for the marooned experimenter in whole-plant physiology. When planning this book it was the complete absence of any comprehensive account, based either statistically or biologically, of the use of fitted curves in plant growth analysis that prompted me to proceed.

In mathematics the problem of giving offence is less acute since the issues involved are for the most part either too simple to warrant critical attention, or else beyond solution.

What of my second guilty admission, the advocacy of empiricism? These waters are too deep to sound here; this I risk in Chapter 3. For the moment let me say that I advocate both empiricism and mechanism in modelling plant growth, but in this place empiricism. One only has to look to commerce and industry, to economics and meteorology, for examples of the great and unashamed use of the approximating function for smoothing, interpolation and prediction. These benefits cannot be denied to students of plant growth because they are supposed to be devoid of mechanistic insight. But, having polarized these two approaches to the study of plant growth it is only fair to draw them some way together again by stating that they are in many respects inseparable, being linked by subterranean channels that are often incompletely understood. Many examples of this will be met with in the course of my text.

What, then, is the book about? It would be too sweeping to say that it surveys the use of fitted regression curves in plant science since it includes no coverage of curves used to describe calibrations, photosynthetic responses, temperature optima, fertilizer responses, allometric relationships, quantal responses (all or none), and the like. By the same token, though when fitting plant growth curves I have inevitably come to favour certain statistical instruments, it would be wrong to confine the text to these alone. Such a course would be too narrow and too suggestive of propaganda. The scope, then, has been defined as one of the middle ways between these two: a review of as complete a selection as possible of the literature in which fitted mathematical functions have been used to link some measure of plant size to the independent variable, time, and used, wherever possible, not merely as representations of the data, but also with some statistical or derivational purpose. By introducing plant growth analysis, the role played in it by the fitted curve, and a sizeable sample of the applications of this approach up to and including 1980, I hope to set the aforementioned reader on the right track for the selection, construction and interpretation of plant growth curves appropriate to his particular needs.

How does the book stand in relation to others in the field? Since these are few an appraisal, albeit partisan, is easy. For many, many years, newcomers to plant growth analysis were served only by a heterogeneous collection of review articles. These naturally increased in number, and varied in approach, as time went by, but it was not until 1972 that G. Clifford Evans's *The Quantitative Analysis of Plant Growth* provided a definitive survey of the field for newcomers and established practitioners alike. So great was its depth of coverage, with extensive excursions into related fields such as experimental design and procedure, environmental measurement and control, and respiratory and photosynthetic studies, that for a time it seemed that little remained to be said on the matter. However, two new considerations gradually arose. One was the need for an introductory text, in book form, for use in relatively short courses on growth analysis, for which *The Quantitative Analysis of Plant Growth,* though eminently readable, was unsuitable for reasons of length alone. The other was the need for some comparative coverage of the use of fitted curves in plant growth analysis, an activity which, while receiving due mention in *The Quantitative Analysis of Plant Growth*, had in the meantime expanded considerably. To meet both of these needs I offered *Plant Growth Analysis* (1978a). Here, within necessarily compact limits, I tried to survey both plant growth analysis itself and the role played in it by fitted curves. Its readers were referred to *The Quantitative Analysis of Plant Growth* for a more thorough exposition of the 'classical' (non-curve-fitting) approach to the subject and are now referred to the following pages for more information on growth curves. In passing, two (quite unrepresentative) comments from reviews of *Plant Growth Analysis* are answered in Chapter 8. Two related volumes remain and one of these, *The Biometry of Plant Growth* (1981) by David R. Causton and Jill C. Venus, is very close in coverage to the present one. Indeed, when the two works were first mooted it was envisaged that they might proceed as a matched pair, uniform in presentation and notation and with a rigid division of coverage. For various reasons, including timing, this proved to be impracticable and so Causton and Venus's book now stands as an advanced and more specialized adjunct to the more introductory and broader treatment of plant growth curves given here. After some duplication between each's introductory chapters, differences soon become apparent to the reader. *The Biometry of Plant Growth* serves the present volume as an extended and rigorous source of information on regression theory and on the use of the Richards function, and on allometry and the analysis of the growth of plant components. In this last respect it has strong links with R. F. Williams's *The Shoot Apex and Leaf Growth* (1975) which, in turn, presents a wealth of morphogenetic observations unmatched by any of the aforementioned.

1

Overture

1.1 Growth

All living organisms are, at various stages in their life history, capable of 'growth' in the sense of change in size, change in form and change in number, given suitable conditions. These three processes together form an important part of the phenomenon of life itself and among natural systems help to distinguish the living from the non-living. Of course, many of the latter may also be said to 'grow': crystals, river deltas and volcanic cones can change recognizably within human time-scales. But, this apart, even within self-reproducing biological organisms a precise definition of what is meant by 'growth' is not at all easy. Definitions may range from an unequivocal statement about change in a specified dimension to a highly abstract state of affairs in which the verb 'to grow' means nothing more than 'to live' or even 'to exist'. Following Hunt (1978a), I advance no firm definition to cover the use of the term in this book other than to say that it will be used to describe irreversible changes with time, mainly in size (however measured), often in form, and occasionally in number.

1.2 Plant growth

From *Biological Abstracts* it is possible to gain the harrowing information that in the last decade alone well over 60 000 separate publications have appeared which may be said to involve, quite specifically, some aspect of 'plant growth'. Clearly, in a work at this level and of this length it is necessary to pare away whole fields of study, from cell division, through growth regulation and morphogenesis, to environmental physiology and agronomy, if what remains is to contribute in a worthwhile way towards filling a gap in the literature. Accordingly, none of the aforementioned topics will receive attention in their own right, the special scope of this work being confined to the analysis by means of fitted curves of series of observations on the growth of plant organs, whole individuals, populations and communities.

What form may such observations take? To introduce the reader to some 'hard data' at the earliest opportunity, I quote a selection from a series of classical experiments performed at Poppelsdorf, West Germany in the 1870s.

In this series, U. Kreusler and his co-workers demonstrated that the growth of an annual plant under natural conditions followed a course that has since been recognized as typical of many. In Table 1.1 their data are given for the increase with time in mean dry weight and leaf area per plant in *Zea mays* (maize) cv. 'Badischer Früh' grown in 1878 (Kreusler, Prehn and Hornberger, 1879). This set of observations was the culmination of several years' work with different varieties of maize and the measurements were heavily replicated. Save for the conversion of cm^2 to m^2, values are reproduced exactly as given by Kreusler.

Table 1.1 Observations made by Kreusler, Prehn and Hornberger (1879) on the growth of 'Badischer Früh' maize at Poppelsdorf in 1878.

Date of harvesting	Day in the year	Mean total dry weight per plant (g)	Mean total leaf area per plant (m^2)
20 May	140	0.3282	n.a.
28 May	148	0.328	n.a.
4 June	155	0.287	n.a.
11 June	162	0.255	0.00179
18 June	169	0.308	0.00292
25 June	176	0.637	0.01244
2 July	183	2.319	0.04192
9 July	190	4.654	0.07622
16 July	197	9.019	0.1301
23 July	204	20.001	0.2136
30 July	211	34.557	0.2805
6 August	218	57.587	0.3384
13 August	225	70.095	0.3047
20 August	232	85.165	0.3025
27 August	239	111.649	0.2976
3 September	246	124.760	0.2684
10 September	253	121.990	0.2387

These primary data are of high quality and remain relevant, even after the passing of more than a century, to the process of describing and interpreting the growth of whole plants; this, despite the fact that they were collected long before modern methods of experimental design and sampling had evolved. Various sets of data from this series have become a classical quarry for those wishing to test suggested improvements in methods of analysis (for references to previous activity of this type see Hunt and Parsons, 1977, Hunt and Evans, 1980 and Parsons and Hunt, 1981). I propose to continue to use these data as a methodological touchstone and they will run like a thread throughout the book, re-appearing in various places and analysed in different ways. What can be seen on first inspection of them?

This particular set forms a series of seventeen observations made, with one exception, at weekly intervals throughout a whole 'growing season'. We see that both measures of plant 'size' span a substantial range; we see also that it was evidently not feasible to determine total leaf area per plant until the fourth sampling occasion. And there we must leave this section, for any more than these preliminary observations on the structure of this data set would come into the category of . . .

1.3 Plant growth analysis

Obvious though it may seem, the first stage in the analysis of data such as those given in Table 1.1 is to plot them out. When this is done (in Fig. 1.1a) we encounter a first difficulty. Because the changes in dry weight over the whole period are of the order of 370-fold, very little of the first phases of development is revealed in this simple plot of dry weight against time on an arithmetic scale. If these data are transformed to (natural) logarithms we can see more clearly what is happening (Fig. 1.1b). There is no special reason why natural, rather than common, logarithms should be used for this purpose (or indeed, if convenient graphical display is the sole objective, some other transformation such as square root), but since these data will be referred to again in contexts that will require this particular transformation, it is convenient to introduce it here.

We see that the plant shows no change in dry weight for the first ten days or so. Then it actually loses weight until about twenty days have passed. Here is an example of another difficulty encountered in defining growth: no increase in weight has occurred but there has been considerable differentiation of leaf tissue in the young seedling (at the expense of total dry weight). From about day 170 the newly-differentiated leaves begin to contribute substantially to carbon assimilation and a so-called grand period of growth begins in

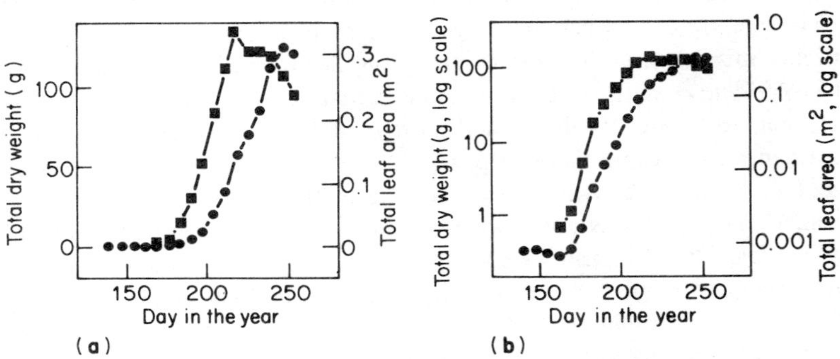

Fig. 1.1 Data from Kreusler *et al.* (1879) on the growth of 'Badischer früh' maize in 1878; (●) mean total dry weight per plant; (■) mean total leaf area per plant. Part **(a)** is plotted on arithmetic and part **(b)** on logarithmic scales. For numerical values see Table 1.1.

which the unfolding of new leaves and an increase in total dry weight occur continuously. The plants flower at about day 205 and an increasing proportion of assimilate is now directed into the developing ear or cob with a corresponding tendency for the lower leaves to atrophy. Finally, at about day 246 net dry weight increase ceases although growth in the sense of a continuing partition of dry weight into ears continues.

This pattern of growth, with great variation in the magnitude of the dry weight values, in the symmetry of the curve and in the time scale which it occupies, is general among annual plants grown in a productive environment. In perennial plants the pattern is similar at first but later, at least in a temperate climate, dry weight increase proceeds in a series of annual steps which may be linked by periods of negative growth in between. Naturally, the environmental conditions affect the magnitude of growth at all stages.

Such preliminary synopses of data can take us only so far, so over the last sixty years an additional body of quantitative techniques has been built up which allows the experimenter to derive important comparative information about the undisturbed growth of whole plants under natural, semi-natural or artificial conditions. These techniques require only the simplest of primary data, such as those described above, and have collectively become known by the informal title 'plant growth analysis'. The origins of this activity have been chronicled by Evans (1972, pp. 189–205) and the whole field is reviewed in Chapter 2, where there are also references to all the alternative résumés. For the time being, it need only be noted that these techniques are, above all, powerful comparative tools, since they have been developed so as to negate, as far as possible, the inherent differences in scale between contrasting organisms so that their performances may be compared on an equable basis. Hunt (1978a, Table 1) listed, as an example of the comparative utility of one of the chief concepts of plant growth analysis, the rates of dry weight increase (irreversible growth in size) for a variety of organisms grown under favourable conditions. These ranged from increases of below 10 per cent per day in large trees, through intermediate rates in herbaceous plants, algae, fungi and microorganisms, to rates exceeding 20 000 per cent per day in an anti-*Escherichia coli* phage. Despite much variation within groups, one broad conclusion was clear: the larger and more complex the organism, the lower the rate of dry weight increase possible, when expressed on a percentage basis. This trend is generally held to be due to the increased morphological and anatomical differentiation which is necessary to sustain life in large systems. This differentiation leads to translocatory pathways of increased length between the point of entry of raw materials into the organism and the site of its nucleoprotein replication (Williams, 1975). The point illustrated was that although the differences in organization between these groups could scarcely be greater, calculations made in this way allowed fair quantitative comparisons to be drawn.

1.4 Plant growth analysis in relation to other fields of activity

Even setting aside the great tracts mentioned in section 1.2, 'plant growth analysis' forms only a part of the activity of studying plant growth in relation to time.

On the broadest scale this study involves an assessment of the primary production of vegetation in the field, considered at the ecosystem and community levels of organization. An outline of the techniques involved here has been given for herbaceous communities by Milner and Hughes (1968) and Singh, Lauenroth and Steinhorst (1975). The role of primary production in the energetics of the whole ecosystem has been summarized by Phillipson (1966). In comparison with studies at this level, the growth analysis of communities focuses attention on much less extensive processes, both in time and in space. Its emphasis is on the more specific and on the more detailed. It has both the disadvantage of providing only a limited, short-term view of events and the advantage of enabling a more precise idea of the nature of plant/time/environment interactions to be gained.

At the population level of organization there coexists the complementary field of demography, on which the study of population dynamics is built. Developed mainly around studies on the human population its aim is more to describe and interpret the changes that occur in numbers of individuals rather than their changes in biomass per individual or their rates of functioning expressed on a unit basis. Keyfitz (1968) has written a mathematical introduction to this field and Smith and Keyfitz (1977) have collected together a very broadly-based anthology of leading publications. In biology, animal science has borrowed more heavily from demography, but there are important implications also for plant science in many of the concepts such as birth and death rates, survivorship and life expectancy (Harper and White, 1974). Solomon (1976) gave an excellent introduction to the biological applications of demography, Sarukhán and Harper (1973) provided a specific botanical example of a study involving some of its techniques, and Harper (1977, pp. 515–643) reviewed progress in the field to that date.

At the organ and organismal levels in plant science demography was not represented until Bazzaz and Harper (1977) published an account of the growth of *Linum usitatissimum* (cultivated flax). These authors applied life-table and other demographic analyses to *leaf* birth and death (in contrast to the more usual case where attention is paid to the birth and death of whole individuals). Hunt (1978b) commented on this new role for demography in relation to the existing role played by plant growth analysis at this level and further work by Hunt and Bazzaz (1980) showed that the leaf demographic approach to the growth of plants operated best when there were many, standard, short-lived 'modules' (clearly identifiable leaves or cohorts of leaves). Plant growth analysis, examining the individual rates of development and functioning of the 'modules', depended on these being relatively fewer, slowly changing and persistent.

Finally, there are studies of plant photosynthetic production – the 'applied' aspects of photosynthesis research – aiming to investigate the plant/environment relationship at the level of the leaf, leaf segment and chloroplast. The various methods appropriate for studies here have been reviewed by Šesták, Čatský and Jarvis (1971). In comparison, plant growth analysis suffers the disadvantage of providing little information about plants to environmental factors, even though valuable clues if not detailed explanations, may sometimes emerge. On the other hand, the great advantage of many of the quantities involved in plant growth analysis is that they provide accurate measurements of the sum performance of the plant integrated both throughout the whole undisturbed plant and across substantial intervals of time. To predict this from the starting point of purely physiological observations would involve many dangerous assumptions. In plant growth analysis the system is judged more by results than by promises. Nátr and Kousalová (1965) and Ondok (1978) also discuss these problems.

So, from the community through to the organ, and beyond, plant growth analysis shares the field with many related activities. In a way, each of these is more specialized than plant growth analysis but none retains its conceptual unity over so wide a range. It remains for the experimenter to examine the possibilities within the various approaches and to enter this continuum at a level appropriate both to the aims of his investigation and to the facilities that he has at hand. If his approach is to involve plant growth analysis, and in particular the use of fitted growth curves, then information in this book will find its mark.

1.5 The 'classical' and the 'functional' approaches to plant growth analysis

A notable dichotomy between two approaches to the subject evolved, in the main, during the 1960s. The above names were first used by Causton (1967); Radford (1967) used the term 'dynamic' for what we shall call the 'functional' approach, but terminology is relatively unimportant provided it is realized that one approach necessarily involves the use of fitted curves and the other does not. We have:

(i) the 'classical' approach, in which the course of events is followed through a series of relatively infrequent, large harvests (with much replication of measurements), the literature in the field running to thousands or tens of thousands of publications;

(ii) the 'functional' approach, in which harvests supplying data for curve-fitting are smaller (less replication of measurements) but more frequent, and of which the publications may be numbered in hundreds.

The two approaches are not mutually exclusive if time and space are no object (harvests may be large *and* frequent) but it is not often that such a scheme makes the most efficient use of the material available. Hence, in most cases, the experimenter must decide in advance which approach to take since this will influence the design and execution of his experiment.

Fortunately, the two approaches share many practical prerequisites, but since these are outside the scope of this book the reader is referred to Hunt (1978a, pp. 5–6) for a brief introduction, or revision, and to Evans (1972, pp. 6–185) for a much fuller survey.

1.6 Computing support required

In the Preface (p. v) I stated that one of the principal objectives of this volume was to point out the opportunities that increasing availability and accessibility of computing power is providing for students of plant growth. While it is certainly possible to scale the lower slopes of the functional approach equipped with nothing more than a set of logarithmic tables, my assumption is that readers will have available some form of computing support without which 'the work involved can easily become impossibly tedious' (Evans, 1972, p. 343).

This support might consist of anything from a programmable pocket calculator to a multimillion-pound institutional computer, so it will obviously be impossible to issue any specific guidance on computational details, particularly in a field which is changing so rapidly. Instead, my strategy will be to outline the biological and statistical advantages and disadvantages of various courses of action, leaving the reader to implement them for himself in his own particular way. The benefits to be gained are such that it is always worth while making a careful assessment of local computing possibilities and, if modest, using them to the full, and in the most advantageous manner.

Smith (1977) has prepared an introduction to scientific analysis on pocket calculators, while for those planning to work with high-level languages on large installations, Nelder (1975) has written a general guidebook which beginners in this, or any other, field of computing will find most valuable. The latter transcends the mere technicalities offered by most computer instruction manuals and covers such vital areas as 'matching the problem to the computer', 'using other people's programs' and 'the organization of data'. Needless to say, all of the activity described in this book is accessible to the well-endowed, either starting from scratch, or using borrowed or library programs. But there may, alas, be places where others will fall by the wayside; in particular where methods involve the use of purpose-written programs in high-level languages. The only two alternatives here are, regrettably, to reconstruct the methods to suit a simpler installation, or to enlist the help of a more fortunate colleague. Despite this, it should not be thought that the field is the exclusive preserve of an elite. As in plant growth analysis itself, much can be done with modest means.

1.7 Units

By international agreement all scientific work is now conducted using the SI system of units (*Système International d'Unités*). The standard unit of length is the metre and of mass the kilogramme. Multiples or fractions of units are restricted to steps of one thousand. For plant growth analysis, this means that we have available millimetres, metres and kilometres for length (and hence area) and microgrammes, milligrammes, grammes and kilogrammes for mass. The very useful centimetre, square centimetre and square decimetre have been sacrificed to a good cause. While this may be regretted, some workers in the field feel that the loss is slight compared to the difficulty that arises when considering volume, where no unit intervenes between the cubic millimetre and the cubic metre, distant by nine powers of ten. The SI units of energy (joule), power (watt) and customary temperature (degree Celsius) can be adopted with few, if any, problems.

The SI unit of time is the second, a unit that may be of greater kinship to physical science than to biology. In plant growth analysis, as in some other fields, the processes studied operate on a longer time scale than this and the everyday units of which days and weeks are by far the most useful to us, have been retained, the only exception being in short-term work where processes are under more or less continuous observation. Incoll, Long and Ashmore (1977) have discussed the particular consequences of SI for experimenters in plant science.

1.8 Notation

In any field of scientific study a consistent notation is a great advantage, especially where equations and mathematical expressions abound. Plant growth analysis had such a fragmentary evolution that none before Evans (1972) attempted a comprehensive and coherent system of notation.

The system that will be followed in this book is, with the unimportant exception of the symbols used for parameters of equations (italic instead of bold lower case), uniform with that of Evans (1972), a policy also followed by Hunt (1978a). As mentioned in the Introduction (p. 4), it was once hoped also to match the system with that of Causton and Venus (1981), but because of the special needs of *The Biometry of Plant Growth*, particularly the requirements of matrix algebra, there had to be some divergence, though many points of overlap remain. The principles to be followed are as follows:

Contractions of names	: small roman captials; e.g. RGR, relative growth rate
Measured quantities	: italic capitals; e.g. W, dry weight
Derived quantities	: bold capitals; e.g. **R**, relative growth rate
Parameters of equations	: italic lower case; e.g. constants a, b
Distinguishing subscripts	: suffix position; e.g. R_W, root dry weight

Subscripts defining time	: prefix position; e.g. ${}_1T$, ${}_1W$, initial time and dry weight
Conventional signs and mathematical symbols	: roman or italic, upper or lower case as required; e.g. W, watt; $\log_e$, natural logarithm; P, probability; t, Student's t.

By way of further example, all of these conventions will have been introduced and used by the end of the next chapter, where special tables will contain synopses of contractions, symbols, expressions, formulae and units. An important point to note is that the measured quantity, time, receives the symbol T and not the more usual t.

It is also important to distinguish between *variates* (quantities having a particular value for each member of a supposedly homogeneous population, with a particular frequency distribution of these values) and *variables* (quantities able to assume different values, but capable of accurate representation on any one occasion). Most measures of plant size come into the former category and time comes into the latter. However, there are some quantities, like temperature, which can be either.

1.9 An outline of the course of the book

Following the disparate collection of preliminaries that appears in this chapter the majority of the book will, as has been stated, be devoted to the use of fitted curves in plant growth analysis. To support this, a general introduction to concepts in plant growth analysis, equally applicable to the classical and the functional approaches, will be given in Chapter 2. Central theoretical issues concerning the rationale behind the use of fitted curves, and the mathematical derivations from them, will be covered in Chapter 3. Chapter 4 will form a practical introduction to the functional approach and Chapters 5 and 6 will review the properties and applications of linear (polynomial) and non-linear functions respectively, with classified guides to existing literature. Various multi-function approaches will be covered in Chapter 7 and Chapter 8 will return to survey the whole field, compare the classical and the functional approaches, and question the future of plant growth analysis in general and the functional approach in particular.

I hope that this structure will allow the reader to skim past areas of which he already has some knowledge, at the same time making it easy for him to relocate these, should the need arise. Since the general objective of this book is to review a marriage between two fields of activity, there will be many places at which the reader can be referred to more specialized literature concerning either one partner or the other. But, as in ecology itself, it is the *relationship* between two pre-existing fields of study that has provided new impetus for scientific advancement and this, I hope will emerge from what follows. This book is not about plant growth; neither is it about regression analysis: it is about the interface of these two.

2

Concepts in plant growth analysis

2.1 Introduction

The purpose of this chapter is to introduce the reader to concepts central to all of plant growth analysis. One great division may immediately be made, and that is according to level of organization. Readers will be familiar with the modern subdivision of biology, not by taxonomic or even by functional criteria, but by organizational structure. This creates a graded series – molecule, organelle, protoplasm, cell, tissue, organ, organism, population, community, ecosystem and biosphere. Concepts in plant growth analysis fit this scheme well since a major historical and motivational division occurs between the organism and below, and the population and above. Yet both parts remain recognizably linked by important similarities, so forming an integrated scheme of activity, accessible to uniform coverage even by a specialized monograph. So, the division between studies concerning plants grown as whole, spaced individuals and plants grown as natural, semi-natural or agricultural populations or communities forms the chief determinant of this chapter's structure. Following a short treatment of generalities shared by all the quantities involved, and of the difference between instantaneous values of these quantities and mean values over a stated interval, concepts in plant growth analysis will be introduced under subheadings according to level of organization. In each, the various quantities will be introduced, defined, assessed and exemplified following a standard pattern; a linkage between the two will also be discussed in section 7.7.

Fundamental to the whole field is a series of sequential measurements of plant size, form or number, the primary data. From these, one or more of four principal types of derived quantity can be constructed:

(i) simple rates of change – rates involving only one variate and time, examples of these being the whole plant's rate of dry weight increase or the rate of increase in numbers of roots per plant;

(ii) simple ratios between two quantities – these may either be ratios between like, such as total leaf dry weight/whole-plant dry weight, or ratios between unlike, such as total leaf area/whole-plant dry weight;

(iii) compounded rates of change – rates involving more than one variate, such as the whole plant's rate of dry weight increase per unit of its leaf area;

(iv) integral durations – estimates of the areas beneath plots of primary or derived quantities and time, such as leaf area duration, leaf area × time.

In addition to identifying many different examples of these four, we shall see a number of important interrelationships between them, one frequent type being (i) = (ii) × (iii).

It is usual to *define* all of these quantities and relationships instantaneously, that is, as they stand at a single point in time. But although this provides an exact exposition, it was for many years necessary to perform *evaluations* of them in the form of means over a stated time interval. If readers remain hazy over the implications of this difference, Hunt's (1978a) analogy may be re-invoked. Here, two ways were suggested in which the speed of a motor car during a journey could be described. One was to take a speedometer reading; this set a more or less instantaneous value ('damping' excepted) to the current speed of the vehicle which, of course, fluctuated widely during most journeys, often changing continuously. The other method was to ignore the fluctuations in speed experienced *en route*, calculating the mean speed over the whole journey from a knowledge of the total distance travelled and the time taken.

This difference in the basis of calculation forms another major division, this time between the classical and the functional approaches to plant growth analysis. The classical approach proceeds to the calculation of mean rates of change, and the like, over periods of time intervening between relatively few harvests (see section 1.5). In the functional approach, mathematical functions are fitted to the primary data, thus describing a relationship between these data and time; from the resulting 'growth curves', fitted values of the data are extracted and are then used to obtain instantaneous values of the various derived quantities. However, it should be remembered that though this difference in approach is important, and strongly evident in the literature, it is possible almost to reverse roles and to use classical methods of analysis to provide very frequent, near-instantaneous values of mean quantities, and to use fitted function for a deliberate smoothing of a non-linear progression in the primary data, in order to obtain a 'forced' mean value for an admittedly variable derived quantity (section 5.2.1).

In passing, it should also be mentioned that only instantaneous values may properly be represented as single points on a progression plotted against time. Mean values should appear as a histogram, with a class interval equal to the harvest interval. This requirement has been ignored rather frequently.

So, bearing in mind that there are, throughout, two chief levels of organization to be considered, four chief types of derived quantity and two methods of calculation, we now need to see an introductory catalogue of the derived quantities before proceeding to consider the theory and practice of curve-fitting. At this point I faced a decision. Either to make Hunt (1978a), or similar, required reading, giving here only the merest outline of the concepts involved; or, to provide a self-contained synopsis of all of plant growth analysis so that the book would be complete in itself. I decided upon the latter course and, drawing on material culled from Chapters 3, 4 and 6 of Hunt (1978a), I have put together a version which has about two thirds of the depth of treatment provided by Hunt (1978a) for the same topics, including some updating. At the end of the chapter there is an annotated list of all alternative synopses, from which further reading may be selected, as necessary.

2.2 Growth analysis of individuals

2.2.1 Introduction

The British School of plant growth analysis crystallized around problems encountered in the 'teens of this century by workers making quantitative assessments of the gross performances of whole plants growing as spaced individuals. The reader has already been referred to Evans (1972) for a detailed history of events at this time, but in order to appreciate the essence of the matter he is invited to consider this problem: two plants have been grown for a week in an experiment; one weighed 1 g at the beginning and one weighed 10 g, at the end of the week it was found that each has increased its weight by 1 g. Which had grown faster?

From one point of view the performances of the two plants are identical since equal amounts of weight have been gained over equal periods of time; in fact, both plants show a 'growth rate' of 1 g week^{-1}. But, armed with the knowledge that their initial weights were so dissimilar it is easy to see that the performance of the lighter plant, which doubled its weight, is in an important sense superior to that of the heavier which increased its weight by only a tenth. Given similar performances during the succeeding weeks, the weight of the heavier plant would soon be equalled by that of the other, which initially was ten times the lighter. Clearly, some measure of growth is needed which takes account of this original difference in size.

2.2.2 Absolute and relative growth rates

The 'growth rate' of the plants in the example given above is an absolute growth rate, **G**. It may be defined as

$$\mathbf{G} = \frac{\mathrm{d}W}{\mathrm{d}T} \cdot \qquad (2.1)$$

G is the instantaneous slope of the plot of total weight per plant, W, against time, T, a plain and simple measure of the rate of increase in weight. The mean absolute growth rate, $\overline{\mathbf{G}}$, over the time interval in question is given by

$$_{1-2}\overline{\mathbf{G}} = \frac{_2W - _1W}{_2T - _1T} , \tag{2.2}$$

where $_1W$ is the weight at time $_1T$, and $_2W$ the weight at time $_2T$. As we have seen, $\overline{\mathbf{G}}$ in both plants had a value of 1 g week^{-1}, so absolute growth rate, AGR, forms a poor comparative tool when we bear in mind the great differences between these two plants in respect of both $_1W$ and $_2W$.

In the financial world, investors, at least in the short term, would inspect the rates of interest earned, not the amounts of capital held, in order to compare financial skills. The measure of plant growth which is analogous to this rate of interest earned is the relative growth rate, RGR. This is the increase in plant material per unit of material per unit of time. In the case of the imaginary experiment described above, the mean RGR of the 1 g plant is 0.693 g g^{-1} week^{-1} and that of the 10 g plant is 0.100 g g^{-1} week^{-1}. At these rates of growth both plants would achieve weights of around 14 g after rather less than four weeks' growth. Obviously, RGR provides a more informative comparison of the plants' relative performances, the more so if it could, by destructive harvesting, be brought to bear on the *dry* weight of the plant. Where did this method of comparison originate and how is it executed?

The direct analogy with financial investment was developed, in the main, by Blackman (1919). In dealing with plant growth he proposed that the rate of interest be termed the 'efficiency index of dry weight production'. This, Blackman held, is

> 'clearly a very important physiological constant. *It represents the efficiency of the plant as a producer of new material*, and gives a measure of the plant's economy in working . . .' (Blackman's italics).

This efficiency index' was defined thus:

$$_2W = {_1W}e^{\mathbf{R}(_2T - _1T)} \tag{2.3}$$

where e is the base of natural logarithms and W is now total *dry* weight per plant. The two exponents which appear on the right-hand side of the equation are **R** (the 'efficiency index') and the time interval itself, $_2T - _1T$. As will be seen, this expression in effect involves a mean value of **R** for this time interval.

West, Briggs and Kidd (1920) suggested the name 'relative growth rate' for **R** and Fisher (1921) pointed out that **R** is most simply expressed as an instantaneous value. In calculus notation this reads

$$\mathbf{R} = \frac{1}{W} \cdot \frac{\mathrm{d}W}{\mathrm{d}T} . \tag{2.4}$$

Another name for **R** is 'specific growth rate'. This term is better, in the sense that it is more in line with modern nomenclature, but it is more recent, less widely used, and not without its own disadvantages (see section 2.2.3).

Fisher's expression for instantaneous relative growth rate (equation 2.4) is an exact notation of the definition given above, namely, the increase in plant weight per unit of plant weight per unit of time. But, although simple in concept, it is not accessible to direct evaluation in a single plant since, according to the rules of calculus (see Machin, 1976, p. 73; Causton, 1977, p. 190) it is equivalent to

$$\mathbf{R} = \frac{\mathrm{d}(\log_e W)}{\mathrm{d}T} . \tag{2.5}$$

This expression tells us that instantaneous relative growth rate, **R**, is the slope of the plot of the natural logarithms of W against T. Most importantly, its value is free to change with different values of T.

In contrast, mean relative growth rate, $\bar{\mathbf{R}}$, is a more involved concept in theoretical terms but a simpler proposition in practice. Mathematically-minded readers will see that it is derived from the integration of equation 2.4 between the limits ${}_1T$ and ${}_2T$, which leads to:

$$\log_e \frac{{}_2W}{{}_1W} = \bar{\mathbf{R}}({}_2T - {}_1T) . \tag{2.6}$$

This may be rearranged as

$${}_{1-2}\bar{\mathbf{R}} = \frac{\log_e {}_2W - \log_e {}_1W}{{}_2T - {}_1T} \tag{2.7}$$

and from here the mean value, ${}_{1-2}\bar{\mathbf{R}}$, may be derived simply by substituting in experimental values of W. As in the case of the car journey, if the amount of change during a given period of time is known then the mean rate of change during this period may be derived.

In practice, the instantaneous values of **R** (equation 2.4) often change smoothly with time and their drift may be followed approximately by deriving values of $\bar{\mathbf{R}}$ for successive harvest intervals via equation 2.7. If the harvest intervals, ${}_2T - {}_1T$, are long $\bar{\mathbf{R}}$ follows **R** only crudely but as the intervals become shorter so the correspondence between these two estimates becomes progressively closer. Hunt (1978a, p. 11) presented a diagram illustrating the derivation of both instantaneous and mean relative growth rates, which may be useful to newcomers to these concepts. For the evaluation of statistical adjuncts to $\bar{\mathbf{R}}$ and related quantities, see Hunt (1978a, p. 12) and Venus and Causton (1979a).

Equation 2.7, when used to calculate values of $\bar{\mathbf{R}}$ for the imaginary example given above, leads, as stated, to values of 0.693 and 0.100 $g\,g^{-1}$ $week^{-1}$ for the smaller and larger plants respectively. The units of $\bar{\mathbf{R}}$ in this case could more economically be expressed as '$week^{-1}$', since the two weight components of the unit cancel into a dimensionless fraction. Of course, any units of the form size $size^{-1}$ $time^{-1}$ may be used but, when dealing with dry weight, as we shall be unless otherwise stated, there are certain advantages attached to each of the more commonly-used versions:

[$g\,g^{-1}$] $week^{-1}$ – the time unit is often the same as the harvest interval and values tend to be of a convenient size;

[$g\,g^{-1}$] day^{-1} – convenient when T is measured in days, but values tend to be rather low;

$mg\,g^{-1}$ $week^{-1}$ – considerable accuracy is available without the need for decimal places.

In addition, it may be mentioned that per cent per week, or per day have also been used where easy interpretation is required across a large range of values, as in section 1.3.

Turning to the real data in Table 1.1 and calculating $\bar{\mathbf{R}}$ for each harvest interval, we can see that in this sample of maize $\bar{\mathbf{R}}_W$ showed an initial fall to below zero (Fig. 2.1). This then recovered, reaching a maximum at about day 180 and then fell to zero again (with some large fluctuations) by day 250. Most of the broader features of this change are an expression of ontogenetic drift. This is due to developments which occur within the plant with the passing of time. But these developments occur against the background of a changing environment (the plants were grown outdoors over the period May to September) and in this case it is impossible to disentangle these internal and external influences on RGR. When plants are grown in a constant, controlled environment for a lengthy period a picture of true ontogenetic drift in RGR emerges. For an example of this in another graminaceous plant, see Fig. 8.6.

Natural courses of relative growth rate such as this have been influenced by virtually all of the environmental variables that have at one time or another been examined. In general, any departure from an adequate supply of light, mineral nutrients or water, or from a suitable temperature regime, or from freedom from external toxins, produces a clearly adverse effect on RGR. Modest increases in carbon dioxide levels or decreases in atmospheric vapour pressure deficits can also promote RGR under certain conditions. Such is the sensitive linkage of this quantity to the whole environmental relations of the plant. It might be added that these factors also interact strongly. For example, the plant's growth response to 'low' levels of one factor depends very much on the available levels of the other factors. The fullest examination yet made of the effects of the environment on the RGR (and other features) of a single

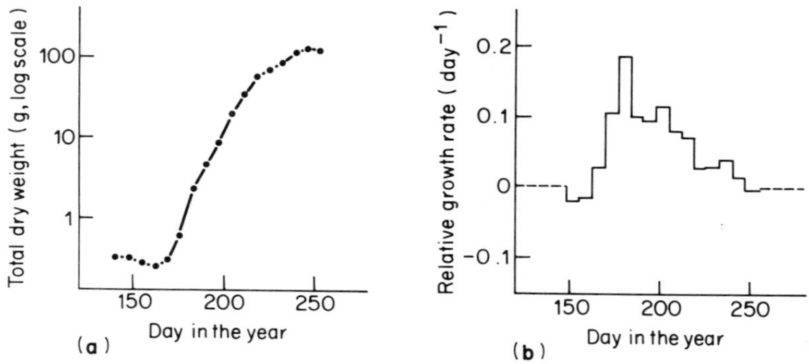

Fig. 2.1 Data from Kreusler *et at.* (1879) on the growth of 'Badischer früh' maize in 1878 (Table 1.1). **(a)** The logarithms of successive mean dry weights joined by straight lines; **(b)** the equivalent progression of mean relative growth rate (equation (2.7).

species under comparable conditions is that of G. C. Evans and A. P. Hughes working with *Impatiens parviflora* (small balsam). This has been summarized by Hughes (1965) and, including later work performed on the same species by other collaborators, by Young (1981).

The first comparison of inter-specific differences in RGR also came from Blackman (1919) who recalculated previously published data on the growth of young crop plants. The subsequent literature contains values of $\mathbf{R}$, or $\bar{\mathbf{R}}$, for several hundred species, but since relatively few of these were grown under strictly comparable conditions only a limited number of comparisons of inter-specific differences is possible. So far, the largest body of comparable data available is that of Grime and Hunt (1975). In this, and in subsequent unpublished work, 144 species (mainly native to Britain) were grown under favourable controlled conditions and their growth was analysed over the period two to five weeks after germination (Fig. 2.2). The maximum instantaneous RGR, $\mathbf{R}_{max}$, observed for each species during this period ranged from 0.22 week^{-1} (in *Picea sitchensis,* Sitka spruce seedlings) to 2.70 week^{-1} (in *Poa annua,* annual poa). The distribution of Grime and Hunt's small samples of woody and annual species were respectively low- and high-biased in $\mathbf{R}_{max}$. Grasses and forbs (herbs that are not grasses) both included a wide range of growth rates.

Intra-specific differences in RGR may occasionally rival the magnitude of inter-specific differences, at least within ecologically similar groups of plants. Within crop species, of course, much comparative information is available from inter-varietal trials, for example those of Duncan and Hesketh (1968). Elias and Chadwick (1979), using virtually the same growing conditions as

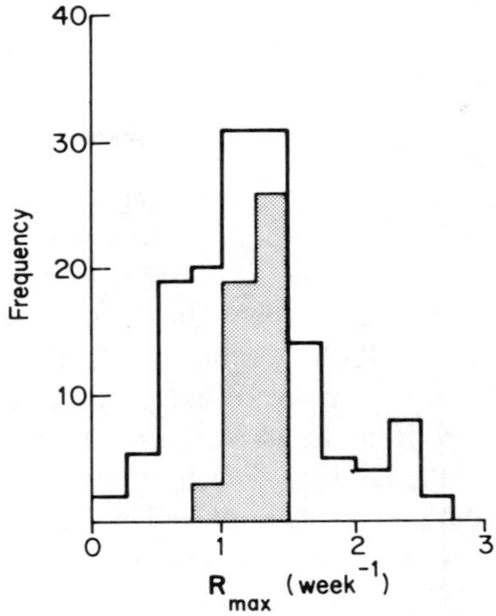

Fig. 2.2 Unshaded area: the distribution of maximum seedling relative growth rate in a sample of 142 species, mainly British natives (data from Grime and Hunt, 1975, supplemented by unpublished material). **Shaded area**: the distribution of mean RGR in 48 cloned subpopulations of *Trifolium repens* (data from Burdon and Harper, 1980, corrected by a factor of × 1.88 to allow for differences in environment and stage of growth). Class intervals of 0.25 week^{-1} are used.

Grime and Hunt (1975), have investigated forty commercially available grass and legume cultivars. They found that inter-varietal differences in $\bar{\mathbf{R}}$ were small. For example, values obtained for *Lolium perenne* (perennial ryegrass) ranged from 1.14 to 1.30 week^{-1} and for *Trifolium repens* (white clover) from 1.09 to 1.30 week^{-1}. Working with clonal material drawn from forty-eight distinct individuals of *Trifolium repens* taken from a single pasture in North Wales, Burdon and Harper (1980) found that, between individuals of this single population, $\bar{\mathbf{R}}$ over four weeks in a glasshouse varied by a factor of 1.8. To put this important result into perspective, data from Burdon and Harper's experiment have been included in Fig. 2.2 after adopting the moderately secure expedient of scaling the data to allow for differences in the two environments and sources of material (cuttings versus seedlings). Grime and Hunt's estimate of $\bar{\mathbf{R}}$ for *Trifolium repens* was 1.26 week^{-1}, compared with Burdon and Harper's grand mean for all clones, 0.67 week^{-1}; values of $\bar{\mathbf{R}}$ brought into Fig. 2.2 from Burdon and Harper's paper are thus 1.88 times the values originally reported. We see that between-individual variation in this species spans three of the eleven class intervals.

On a broader geographical scale, Eagles (1969) has shown that in a productive, controlled environment there can often be a two-fold difference in

R between Norwegian and Portuguese populations of *Dactylis glomerata* (cocksfoot) although both the direction and magnitude of this differential are subject to marked interactions of ontogeny and environment.

2.2.3 Other simple rates

Relative growth rate provides a convenient integration of the combined performances of the various parts of the plant. It is especially useful when the need arises to compare species and treatment differences on a uniform basis. But when calculated at the whole plant level it tells us nothing of the causal processes which contribute to the plant's gross performance. One further step in this direction is possible before going on to examine other ways of assessing the growth of the plant. This is to calculate RGR for each subcomponent of the plant, say root and shoot; or leaves, stem, and root; or leaves of varying ages, petioles, stem, main and lateral roots. The subdivisions are dictated only by convenience and the computations are directly analogous to those for whole plant RGR. Similarly, important new information may be gained from calculations made on the basis of fresh weight, volume, area, length or number, or on the basis of the content of various metabolic compounds such as selected carbohydrates or proteins, or even the calorific content of the plant material.

In this wider context the term 'relative growth rate' is to be preferred to 'specific growth rate' since the latter refers exclusively to growth in relation to mass. So, even though *undefined* RGR should always be interpreted as being $(1/W)(\mathrm{d}W/\mathrm{d}T)$, this does not preclude the term from being used in a generic sense: suitably defined, its scope can be widened considerably.

All of the foregoing possibilities apply equally to absolute growth rate, with the caveat throughout that AGR includes no correction for the size of the system under investigation, indicating only in absolute terms the extent of its change in size.

Hunt (1978a, p. 58) covered the relationship between RGR, doubling time and half-life, quoting the identity

$$\text{Doubling time} = \frac{0.693}{\mathbf{R}} \ . \qquad (2.8)$$

If the instantaneous relative growth rate, **R**, is used in the above equation then the instantaneous doubling time will be derived; if the mean relative growth rate, $\bar{\mathbf{R}}$, is used then the mean time for doubling, estimated over the same time interval, will result. As we have seen, the units of **R** (or $\bar{\mathbf{R}}$) are time^{-1} and so, as expected, the units of doubling time will simply be time alone. In situations where **R** or $\bar{\mathbf{R}}$ has a negative value, the system under investigation is not growing but decaying. Solving equation 2.8 in such cases would provide not a doubling time but a 'halving time', more commonly referred to as half-life and discussed further in section 5.2.1.

2.2.4 Unit leaf rate and leaf area ratio

The expression for instantaneous RGR

$$\mathbf{R} = \frac{1}{W} \cdot \frac{\mathrm{d}W}{\mathrm{d}T} \tag{2.4}$$

treats all of the weight of the plant as being equally productive of further weight. We know that as plants grow the proportion of purely structural material that they contain increases, for much the same reasons that larger animals develop proportionately more bulky bones than smaller ones (Alexander, 1971, p. 53). So this implied notion of mere weight being all that is necessary to provide still more weight becomes more and more improbable as growth proceeds. A consequence of this is the ontogenetic decline in RGR seen in Figs 2.1 and 8.6. What is needed is an index of the productive efficiency of plants in relation to some clearly-identifiable component of relatively constant performance. Gregory (1918) suggested that the net gain in weight per unit of leaf area (the 'average rate of assimilation') might be this more meaningful index of growth. It is a 'type (iii)' quantity (see section 2.1). Briggs, Kidd and West (1920) termed it 'unit leaf rate' (ULR). This quantity is conventionally given the symbol **E** and the expression for its instantaneous value is

$$\mathbf{E} = \frac{1}{L_\mathrm{A}} \cdot \frac{\mathrm{d}W}{\mathrm{d}T} \tag{2.9}$$

where L_A is the total leaf area present on the plant. Williams (1946) provided a convenient formula for estimating mean ULR, $\bar{\mathbf{E}}$, over a period of time:

$${}_{1\text{-}2}\bar{\mathbf{E}} = \frac{{}_2W - {}_1W}{{}_2T - {}_1T} \cdot \frac{\log_\mathrm{e} {}_2L_\mathrm{A} - \log_\mathrm{e} {}_1L_\mathrm{A}}{{}_2L_\mathrm{A} - {}_1L_\mathrm{A}} \; . \tag{2.10}$$

Unit leaf rate has also widely been called 'net assimilation rate', NAR, (Gregory, 1926) but the former term is both older and more suitable, as Evans (1972, p. 205) explained.

Armed with what might, but for the inclusion of the weight of mineral elements in W, be an estimate of the carbon-assimilatory capacity of the leaves, all one now lacks is an estimate of the leafiness of the plant before being in a position to calculate the overall relative growth rate. Alternatively, beginning with RGR, this index of leafiness is the other quantity that can be derived, with ULR, to produce an informatively subdivided summary of the plant's performance. Briggs, Kidd and West (1920) called this other quantity the leaf area ratio (LAR) and defined it as the ratio of total leaf area to whole plant dry weight, a 'type (ii)' quantity (see section 2.1). It can be notated as **F**:

$$\mathbf{F} = \frac{L_\mathrm{A}}{W} \; . \tag{2.11}$$

In a broad sense, LAR represents the ratio of photosynthesizing to respiring material within the plant.

Over a harvest interval its mean value, $\bar{\mathbf{F}}$, is simply given by

$$ {}_{1-2}\bar{\mathbf{F}} = \frac{({}_1L_A/{}_1W) + ({}_2L_A/{}_2W)}{2} = \frac{{}_1\mathbf{F} + {}_2\mathbf{F}}{2} \qquad (2.12) $$

if one assumes that **F** is linearly related to time (Ondok, 1971a, discusses other cases). Equation 2.11 supplies a straightforward estimate of instantaneous value using corresponding values of L_A and W from a single harvest.

Since ULR and LAR evolved simultaneously as subdivisions of RGR it is by definition that

$$ \mathbf{R} = \mathbf{E} \times \mathbf{F} \qquad (2.13) $$

since

$$ \frac{1}{W} \cdot \frac{dW}{dT} = \frac{1}{L_A} \cdot \frac{dW}{dT} \times \frac{L_A}{W} \; . \qquad (2.14) $$

Simply expressed, the growth rate of the plant depends simultaneously upon the efficiency of its leaves as producers of new material and upon the leafiness of the plant itself. But, except in very special circumstances,

$$ \bar{\mathbf{R}} \neq \bar{\mathbf{E}} \times \bar{\mathbf{F}} \qquad (2.15) $$

because equation 2.13 holds only crudely for mean values of the three quantities. Instantaneous values are needed for this relationship to be precise. There are also other assumptions involved in the use of equation 2.10 from which the experimenter following the classical approach can run into difficulty if either (a) his plants are growing quickly or (b) his harvest intervals are long (Evans and Hughes, 1962; Whitehead and Myerscough, 1962). Evans (1972, p. 268) discusses this problem and points out the ways in which it may be overcome by further computation.

One of the factors that led to the introduction of unit leaf rate was a search for a relatively constant index of growth that was independent of plant size. Although, in general, ULR proved to be stable for longer periods than RGR, ontogenetic drift was still detected. Can we see this if we return to Kreusler's data for maize (Table 1.1)? Figure 2.3 presents values for $\bar{\mathbf{F}}$ and $\bar{\mathbf{E}}$ calculated by way of equations 2.12 and 2.10. $\bar{\mathbf{F}}$ quickly reaches a maximum around day 180 then declines smoothly to a low value: the crop's leafiness is maximized very early. $\bar{\mathbf{E}}$, apart from a sharp rise and fall at the beginning and end lies mainly in the range 5–10 g m^{-2} day^{-1}. Great variability is exhibited, a noted feature of the classical approach (see section 2.2), which prevents, on this occasion, any firm assessment of the reality or not of ontogenetic drift (see Hunt and Evans (1980) for further discussion of trends in LAR and ULR in these data).

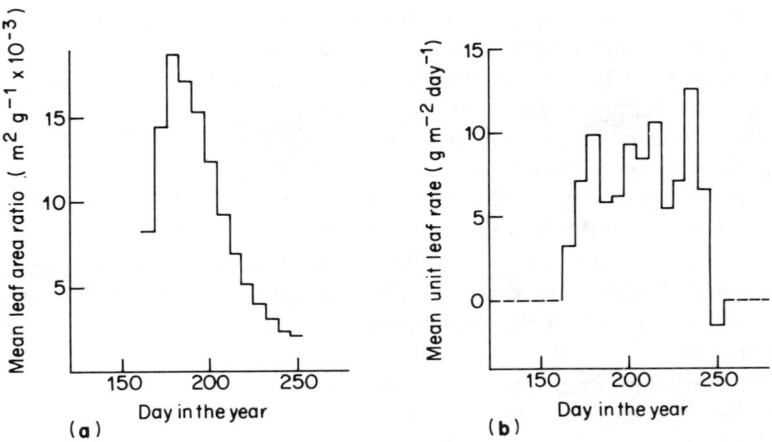

Fig. 2.3 Data from Kreusler *et al.* (1879) on the growth of 'Badischer früh' maize in 1878 (Table 1.1). **(a)** Mean values of leaf area ratio calculated for each harvest interval (equation 2.12); **(b)** mean values of unit leaf rate calculated for each harvest interval (equation 2.10).

Williams (1946) showed that for annual plants grown in a constant environment, the closer the approach to an effective measure of assimilating capacity, the more reliable and characteristic of the species become the estimates of $\bar{\mathbf{E}}$ derived from this measure. When calculated on the bases of leaf weight, leaf area and leaf protein a clear series of increasing stability of $\bar{\mathbf{E}}$ emerged, particularly when nitrogen was in relatively short supply. These three versions of $\bar{\mathbf{E}}$ may be notated $\bar{\mathbf{E}}_W$, $\bar{\mathbf{E}}_A$ and $\bar{\mathbf{E}}_P$ respectively. In cases where it is not feasible to estimate, $\bar{\mathbf{E}}_P$, $\bar{\mathbf{E}}_A$ can usually provide an acceptable substitute.

In addition, it may be mentioned that some workers have calculated ULR not on the basis of L_A but on total chlorophyll content of the plant. This alternative measure of the size of the plant's assimilatory apparatus is held to be most valuable when plants are light-limited (Květ *et al.*, 1971). ULR calculated in this way is notated $\mathbf{E}_C$.

In a simultaneous comparison of the growth of *Hordeum vulgare* (barley) outdoors at Ottawa and in a controlled environment, Thorne (1961) showed that a downward ontogenetic drift in $\bar{\mathbf{E}}_A$ was accentuated by the declining favourability of outdoor conditions in the late summer. A striking example of this dependence of $\bar{\mathbf{E}}_A$ on the external environment was provided by the work of Watson (1947). Each of four species, *Beta vulgaris* (sugar beet), *Hordeum vulgare* (barley), *Solanum tuberosum* (potato) and *Triticum* sp. (wheat) grown in a field at Rothamsted, showed a peak in $\bar{\mathbf{E}}_A$ near the summer solstice (late June) irrespective of the nature of the crop or of the time of planting. These peaks were principally the result of a combination of high solar insolation, high temperature and long day-length.

Each of these factors may be investigated singly under controlled or semi-controlled conditions. For example, Blackman and Wilson (1951) used a series of shade screens to demonstrate the dependence of $\bar{\mathbf{R}}$, $\bar{\mathbf{E}}_A$ and $\bar{\mathbf{F}}$ upon the level of illumination received. In general, $\bar{\mathbf{E}}_A$ was found to be related linearly and positively to the logarithm of percentage full daylight; in $\bar{\mathbf{F}}$ this relationship was a negative one. The slopes of these two relationships determined the trend obtained for $\bar{\mathbf{R}}$. In one extreme case, *Geum urbanum* (wood avens), maximum $\bar{\mathbf{R}}$ was predicted to occur at only 54 per cent of full daylight because of the steep plunge in $\bar{\mathbf{F}}$ that occurred with increasing light intensity.

Wide variation in $\bar{\mathbf{E}}_A$ may occur between species. For example, unpublished data from the experiments of Grime and Hunt (1975) showed that values of $\mathbf{E}_A$ (instantaneous maxima, $\mathbf{E}_{max}$) varied from 19.7 g m^{-2} week^{-1} (*Vaccinium vitis-idaea*, cowberry) to 192 g m^{-2} week^{-1} (*Cerastium holosteoides*, mouse-ear chickweed) in this particular sample of species. Dicotyledons, monocotyledons and annual species showed no distinct bias towards high or low $\mathbf{E}_{max}$ but woody species were clearly low-biased. This last observation supports that of Coombe (1960), who showed that woody plants exhibited inherently lower $\mathbf{E}_A$ than the majority of herbaceous species, a conclusion also reached by Jarvis and Jarvis (1964). These workers grew a variety of coniferous species and sunflower under productive, controlled conditions. They showed that in comparison with sunflower, the lower $\bar{\mathbf{R}}$ of the conifer seedlings was due more to a low $\bar{\mathbf{F}}$ than to a low $\bar{\mathbf{E}}_A$ (although there were important differences in $\bar{\mathbf{E}}_A$ also). Table 2.1 gives a typical comparison (equation 2.15 is in operation here).

Table 2.1 Contribution of $\bar{\mathbf{E}}_A$ and $\bar{\mathbf{F}}$ to seedling $\bar{\mathbf{R}}$. (Recalculated from data given by Jarvis and Jarvis, 1964.)

Species and harvest interval	$\bar{\mathbf{R}}$ (week^{-1})	$\bar{\mathbf{E}}_A$ (g m^{-2} week^{-1})	$\bar{\mathbf{F}}$ (m^2 g^{-1})
Pinus sylvestris (Scots pine) 42 days	0.135	32.3	0.0044
Helianthus annuus (sunflower) 8 days	0.955	59.3	0.0177

2.2.5 Specific leaf area and leaf weight ratio

If leaf dry weight, L_W, is known a useful subdivision of LAR into two further 'type (ii)' quantities may be made:

$$\frac{L_A}{W} = \frac{L_A}{L_W} \times \frac{L_W}{W} \tag{2.16}$$

where L_A/L_W is the specific leaf area, SLA, the mean area of leaf displayed per unit of leaf weight (in a sense a measure of leaf density or relative thickness). L_W/W is the leaf weight ratio, LWR, a dimensionless index of the leafiness of the plant on a weight basis (cf. SLA, the leafiness of the plant on an area/weight basis). These subdivisions of LAR may be inserted into equation 2.14 to give:

$$\frac{1}{W} \cdot \frac{dW}{dT} = \frac{1}{L_A} \cdot \frac{dW}{dT} \times \frac{L_A}{L_W} \times \frac{L_W}{W} \quad . \tag{2.17}$$

No single symbols are commonly in use either for SLA or for LWR. As with LAR, both are amenable to calculation of instantaneous values at the time of harvest. Mean values between harvest intervals may also be estimated in the same manner as those of LAR (equation 2.12).

Of the two, SLA and LWR, the former is in general both the more sensitive to environmental change and the more prone to ontogenetic drift. Referring to the growth of *Impatiens parviflora* (small balsam) Evans (1972, p. 331) wrote:

> 'For leaf weight we could draw up a list of environmental factors, changes in which hardly affected the value of LWR. It would be profitless to attempt to do the same for specific leaf area, as there has been at least some influence of the environment in every instance which we have examined.'

Foremost in this second list might be variation in light intensity: deep shade causes striking increases in SLA, partly offsetting decreases in ULR (Hughes and Evans, 1962).

Differences in both LWR and SLA occur even between closely-related species. For example, Evans (1972, p. 442) explained that in two species of sunflower,

> '*Helianthus debilis* has a substantially higher proportion of its dry matter in the form of leaves than has *H. annuus*';

values of LWR were commonly greater by *c.*20 per cent. Jarvis and Jarvis (1964) established that the much less leafy nature of Scots pine, in comparison with sunflower (Table 2.1), was due almost entirely to the relatively greater density of the pine needles and hardly at all to variation in LWR (the 'productive investment' of the plant) which, in fact, showed a small difference in favour of the pine (Table 2.2).

2.2.6 Other simple ratios

In the growth analysis of the individual plant the number of possible 'type (ii)'

Table 2.2 Contribution of SLA and LWR to seedling LAR (instantaneous values). (Recalculated from data given by Jarvis and Jarvis, 1964.) See also Table 2.1.

Species and harvest	LAR ($m^2 g^{-1}$)	SLA ($m^2 g^{-1}$)	LWR (dimensionless)
Pinus sylvestris, Scots pine (at 2 years)	0.0054	0.0084	0.643
Helianthus annuus, sunflower (at *c.* 100 mm height)	0.0234	0.0432	0.542

ratios is almost limitless. We have already encountered LAR, SLA and LWR and seen that in the latter case, where both parts of the fraction bear the same units, the quantity is a simple index of the importance of one component of the plant in relation to the whole. Other ratios of this general type are R_W/W, the root weight ratio, where R_W is the total dry weight of roots on the plant, and its associate, the shoot weight ratio, S_W/W, where S_W is the total above-ground dry weight of the plant. For an example of broad intra- and inter-specific comparisons of this type of quantity see Elias and Chadwick (1979). These authors derived separate root, stem and leaf weight ratios, which are related by the trivial expression

$$\frac{L_W}{W} + \frac{Stem_W}{W} + \frac{R_W}{W} = 1 \cdot \qquad (2.18)$$

making a valuable series of comparisons between their forty species and cultivars.

It is possible to accentuate top-root relationships by calculating the root-shoot ratio, R_W/S_W, or its more logical, but in pronunciation less negotiable, converse the shoot-root ratio, S_W/R_W. The advantage here is that the quantity is, by its very structure, more responsive to change than the previous root or shoot weight ratios, particularly when $S_W \gg R_W$, or *vice versa*. In some cases this increased responsiveness can border on instability and so care is needed. Nonetheless, the quantity has been very popular because it is so sensitive to environmental influences (see Hunt and Burnett (1973) for references).

The fresh weight-dry weight ratio, FW/W, where FW is the whole plant fresh (wet) weight, is often a useful index of the water content of the plant. Similarly a mineral nutrient-dry weight ratio, nothing more than a gravimetric nutrient concentration, is also frequently used.

In all of these, direct calculations from primary data are feasible in each case and mean values over a harvest interval may be estimated using methods analogous to those for $\bar{\mathbf{F}}$ (equation 2.12). All ratios are subject to genetic, ontogenetic and environmental control.

2.2.7 Other compounded rates

If Y and Z are symbols representing any of the primary data in plant growth analysis (the various weights, areas, volumes or numbers) then 'type (iii)' quantities may be given the general notation $(1/Z)(\mathrm{d}Y/\mathrm{d}T)$. In plain words, these are 'rates of production of something per unit of something else'. In plant growth analysis, provided that the 'something', Y, is of interest to the experimenter and that the 'something else', Z, may reasonably be held responsible for its production, then $(1/Z)(\mathrm{d}Y/\mathrm{d}T)$ is an analytical tool of fundamental importance. The following paragraphs describe the use of this tool in a selection of guises. The list is not exhaustive, nor by any means have all of the possibilities in this direction been followed up, so the reader may well be stimulated to invent a few of his own. All will be defined instantaneously, but are accessible to evaluation as harvest-interval means via equation 2.10.

When discussing unit leaf rate, $\bar{\mathbf{E}}$, in section 2.2.4 we saw that it could be calculated on the basis of leaf weight ($\mathbf{E_W}$), leaf area ($\mathbf{E_A}$) or leaf protein content ($\mathbf{E_P}$). This comprised an increasingly informative series for the experimenter interested in the detailed functioning of the plant, but at the cost of increased experimental labour. Hunt and Burnett (1973) devised a unit shoot rate (USR), the rate of production of dry weight per unit of shoot material, a quantity termed 'net top effectiveness' by Barrow (1975). This may be notated as $\mathbf{B}$ (Hunt, 1978a) and defined as an instantaneous value.

$$\mathbf{B} = \frac{1}{S_\mathrm{W}} \cdot \frac{\mathrm{d}W}{\mathrm{d}T} \tag{2.19}$$

where, as before, S_W is the total dry weight of the above-ground parts of the plant and W is the dry weight of the whole plant. Unit leaf rate may be regarded as a component of unit shoot rate since

$$\frac{1}{S_\mathrm{W}} \cdot \frac{\mathrm{d}W}{\mathrm{d}T} = \frac{1}{L_\mathrm{A}} \cdot \frac{\mathrm{d}W}{\mathrm{d}T} \times \frac{L_\mathrm{A}}{S_\mathrm{W}} \tag{2.20}$$

or

$$\mathbf{B} = \mathbf{E} \times \frac{L_\mathrm{A}}{S_\mathrm{W}} \,. \tag{2.21}$$

Even more than $\mathbf{E_W}$, $\mathbf{B}$ provides a crude but simple index of the performance of the productive parts of the plant. The series $\mathbf{B}$, $\mathbf{E_W}$, $\mathbf{E_A}$, $\mathbf{E_P}$ is thus one of increasing sophistication but decreasing ease of derivation. $\mathbf{B}$ is worth considering in two cases: (a) where it is difficult or unduly labourious to record L_W separately from S_W, and (b) where there is little point in making this distinction (e.g. in young monocotyledons where leaves constitute a huge proportion of the above ground parts of the plant).

Williams (1948) commented that, in detailed studies on the mineral nutrition of plants, simple calculations yielding rates of uptake of mineral nutrients per day were complicated by what he called the 'size factor of the absorbing system'. He suggested that comparisons might be made on an equable basis if a concept analogous to unit leaf rate were to be employed. He defined

$$\text{Rate of intake} = \frac{1}{R_W} \cdot \frac{dM}{dT} \tag{2.22}$$

where M is the plant's content of the mineral under consideration and R_W is again the dry weight of the root system. The whole quantity provides an estimate of the rate of nutrient uptake per unit weight of root, a sort of root efficiency (Hunt, 1973). Welbank (1962) suggested the name specific absorption rate (SAR) for Williams's 'rate of intake', and used the symbol **A**. An example of SAR in operation may be drawn from the work of Welbank (1964). *Beta vulgaris* (sugar beet) was grown in soil in pots outdoors at Rothamsted. Three levels of added nitrogen and three of phosphorus were combined factorially and plants were grown both with and without competition from *Agropyron repens* (couch grass). At all levels of added nutrients, competition from *Agropyron* depressed the crop's mean specific absorption rate for nitrogen substantially ($\bar{\mathbf{A}}_N$, root dry weight basis). The concept of SAR on a root weight basis is informative only for so long as R_W 'may reasonably be held responsible' for the intake of M. Perhaps root length, area, volume or number would be better? Hackett (1969) and Evans (1972, p. 228) discuss these problems further.

Using arguments analogous to those of Williams (1948), Keay, Biddiscombe and Ozanne (1970) suggested that analyses of the utilization of mineral nutrients in plants, sometimes expressed as the reciprocal of the concentration of nutrients in the dry matter, would better describe the pattern with time in this utilization if expressed as the rate of dry weight increment per unit of absorbed nutrient. Unfortunately, Keay, Biddiscombe and Ozanne also used the symbol A for this quantity, conflicting with the **A** of Welbank (1962). Hunt (1978a) suggested the term 'specific utilization rate' (SUR), and the symbol **U**, so that

$$\mathbf{U} = \frac{1}{M} \cdot \frac{dW}{dT} \ . \tag{2.23}$$

SUR is almost the converse of specific absorption rate since

$$\frac{1}{R_W} \cdot \frac{dM}{dT} \times \frac{1}{M} \cdot \frac{dW}{dT} = \frac{1}{M} \cdot \frac{dM}{dT} \times \frac{1}{W} \cdot \frac{dW}{dT} \times \frac{W}{R_W} \tag{2.24}$$

or

$$\mathbf{A} \times \mathbf{U} = \mathbf{R_M} \times \mathbf{R_W} \times \frac{W}{R_W} \qquad (2.25)$$

where $\mathbf{R_M}$ and $\mathbf{R_W}$ are the relative growth rates in mineral nutrient content and in whole plant dry weight respectively. W/R_W is, of course, the reciprocal of the root weight ratio (section 2.2.6). As constructed, **SUR** is a measure of the efficiency with which dry weight is increased by mineral uptake, although clearly this concept needs to be handled with care since the causal connection between the two processes is not rigid.

In a study of the dynamics of leaf growth in *Trifolium subterraneum* (subterranean clover) Williams (1975, p. 175) developed a general-purpose rate of production for one subcellular component of the leaf per unit of another component. This he termed **G**, and appended appropriate subscripts to describe the particular rate under examination. These subscripts serve to distinguish it from absolute growth rate, **G**. One example of such a rate is

$$\mathbf{G}_{\mathbf{PN,RNA}} = \frac{1}{RNA} \cdot \frac{\mathrm{d}PN}{\mathrm{d}T} \quad . \qquad (2.26)$$

This represents the instantaneous rate of production of protein nitrogen, *PN*, per unit weight of ribonucleic acid, *RNA*. Williams was able to show that in young clover leaves this functional efficiency of *RNA* as a producer of *PN* declined markedly with time – a valuable addition to the concept of ontogenetic drift. He surmised that since values of $\bar{\mathbf{G}}_{\mathbf{PN,RNA}}$ fell roughly 'into three groups, with mean values of 5.6, 4.4 and 1.5 day^{-1} for days 10–13, 13–19 and 19–25 respectively', this indicated that 'the synthesis of specific proteins or groups of proteins might be dominant for successive stages of development.'

When analysing the growth of perennial crops which lose part of their accumulated dry weight during the winter, for example deciduous trees, a difficulty arises in applying the concept of relative growth rate since temporary summer biomass interferes with an exact measurement of the functional efficiency of the plant's dry weight as a producer of new material. To overcome this problem, Dudney (1973) distinguished the accumulated current mass of the crop, *W*, from the mass of the perennating structure, which we can designate, W_P. He devised a 'unit production rate', **UPR**, allocating the symbol Π (pi), such that

$$\Pi = \frac{1}{W_P} \cdot \frac{\mathrm{d}W}{\mathrm{d}T} \quad . \qquad (2.27)$$

This provided a measure of the efficiency with which the perennating mass of the tree provided further dry weight. Dudney (1974) found that $\mathbf{\Pi}$ was greater in spur-pruned apple trees than in minimal-pruned apple trees for the first five years following the introduction of their respective pruning regimes; thereafter this difference was reversed. Throughout, **Π** declined with age, confirming for a perennial plant what we have already seen in an annual one (Fig. 2.1).

Following suggestions first made by Causton (1967) and developed by Hunt (1978b), Hunt and Bazzaz (1980) subdivided the whole-plant's relative growth rate in weight, $\mathbf{R}_W$, into an expression including the relative growth rates of the component parts of the plant. Hence:

$$\mathbf{R}_W = \frac{1}{W}\cdot\frac{dW}{dT} = \left(\frac{1}{w_1}\cdot\frac{dw_1}{dT}\right)\cdot\frac{w_1}{W} + \left(\frac{1}{w_2}\cdot\frac{dw_2}{dT}\right)\cdot\frac{w_2}{W}\ldots + \left(\frac{1}{w_n}\cdot\frac{dw_n}{dT}\right)\cdot\frac{w_n}{W} \tag{2.28}$$

where $w_1, w_2, \ldots w_n$ are the dry weights of the individual parts of the plant, such as roots, stems and the successive leaves, in sum equalling W. Each of the terms on the right-hand side of equation 2.28 consists of the product of, firstly, the appropriate part's relative growth rate and, secondly, the ratio of that part to the whole, that is,

$$\mathbf{R}_W = \mathbf{R}_1\cdot\frac{w_1}{W} + \mathbf{R}_2\cdot\frac{w_2}{W}\ldots + \mathbf{R}_n\cdot\frac{w_n}{W} \tag{2.29}$$

where $\mathbf{R}_1, \mathbf{R}_2, \ldots \mathbf{R}_n$ are the relative growth rates of $w_1, w_2, \ldots w_n$. In practice, it was found to be more convenient to evaluate each of the right-hand terms in equation 2.29 as single quantities like $(1/W)(dw_1/dT)$. This is the rate of production of the component w_1 per unit of whole plant dry weight, an index of the current commitment of the plant to the production of this particular part which also takes into account the current size of the whole plant. Hunt and Bazzaz named such quantities 'component production rates', allocating to them the symbol **J**. Component production rates sum up to the whole-plant relative growth rate, that is:

$$\mathbf{R}_W = \mathbf{J}_1 + \mathbf{J}_2 \ldots + \mathbf{J}_n \tag{2.30}$$

where 1 . . . n represent the separate components, as before. Hunt and Bazzaz applied this analytical technique to data on the growth of individual nodal clusters of leaves of *Ambrosia trifida* (giant ragweed) obtained from a glasshouse experiment performed at Urbana, Illinois, U.S.A. CPRs for leaves at nodes 2, 3 and 4 were around $0.04\,g\,g^{-1}\,day^{-1}$ in the more fertile of two treatments: about twice the values obtained for the same leaves in the less fertile treatment.

So, the preceding six paragraphs illustrate how quantities of the form $(1/Z)(dY/dT)$ can aid our understanding of processes from the molecular level ($\mathbf{G}_{PN,RNA}$), through organs (**J**), and whole herbaceous plants (**A**, **B**, **U**) to perennial woody crops (**Π**). Provided that the caveat contained in the first paragraph of this section is borne in mind, there is no reason why applications of this concept should not flourish still further.

2.3 Growth analysis of populations and communities

2.3.1 Links with the growth analysis of individuals

All of the topics dealt with in section 2.2 concern the growth of plants as spaced individuals. This is not to say that the various concepts were applied only to plants grown singly; far from it, since large populations of similar individuals normally need to be raised to meet the demands of sequential destructive harvesting. Rather that the results, once obtained, were expressed on a 'per plant' basis. The size of the population, and the sum performance of its constituents, were not matters of concern in themselves.

However, in agriculture particularly, and in some studies of production in natural vegetation, it has often been informative to treat both populations (such as monospecific crop stands) and communities (such as mixed-species grassland) as single functional units, expressing their overall performances in terms parallel to those used in the growth analysis of individual plants. It should be emphasized that there is no theoretical reason why the concepts of relative growth rate, unit leaf rate, leaf area ratio and so on, should not themselves be applied on a 'per crop' basis instead of on a 'per individual' basis. But, in practice, this parallel series of analytical procedures has borrowed only one of these, ULR, for use in an unaltered form. Other concepts, designed exclusively for the study of population and community growth, have been developed to supplement it. These will be discussed in their own right in the first instance, before considering a synthesis of them with the concepts involved in the growth analysis of individual plants (section 7.7).

2.3.2 Leaf area index

In the study of population and community growth, ULR provides exactly the same information as it does in the study of the growth of individuals, namely, an index of the functional efficiency of the productive parts of the plant. Given this, a knowledge of the leafiness of the crop is also needed before its full performance can be assessed. Leaf area per plant is an inappropriate measure of the leafiness of a whole crop since it takes no account of the spacing of the plants, a factor which must clearly be involved in any estimate of 'crop leafiness'. To overcome this difficulty Watson (1947) introduced the more crop-orientated concept of leafiness in relation to land area. This he named leaf area index, LAI, and defined it as leaf area per unit area of land. Using P to represent the land area and L_A to represent, not the total leaf area per plant as previously, but the total leaf area above the land area P, leaf area index may be given the single symbol $\mathbf{L}$ and notated

$$\mathbf{L} = \frac{L_A}{P} \quad . \tag{2.31}$$

LAI is the functional size of the crop standing on the land area P.

If the dimensions of L_A and P are both the same (e.g. area) then L itself has no units: it is a dimensionless ratio and falls within 'type (ii)'. To determine instantaneous values of LAI, $\mathbf{L}$, requires only that L_A be measured in an adequate and representative number of small samples of the crop, each from a known land area, P. Alternatively, L_A may be estimated on a per plant basis and then multiplied by the measured current plant density in the crop. Mean values can be calculated analogously to those of LAR (equation 2.12).

In effect, leaf area index is the number of complete layers of leaves displayed by the crop, expressed as an average for the whole crop. This concept is inevitably a crude one since leaves never actually form complete unbroken layers arranged one above the other. Leaves are commonly displayed at varying angles to the horizontal and these angles vary with the morphology of the species and with the conditions under which it is being grown. LAI is only an index of *mean* crop leafiness since, even at relatively high values of LAI, random holes occur in the canopy where there is no leaf cover at all. Moreover, leaves of different layers in the canopy experience different environmental conditions and function in different ways. Nevertheless, as a broad index of the productive capacity of a stand of vegetation LAI has been of considerable value.

In a newly-germinated crop $\mathbf{L}$ remains below 1.0 for some time, since the total leaf area of the young seedlings is negligible in relation to the land area on which they stand. As the crop develops $\mathbf{L}$ increases, until it reaches its maximum value (often in the range 2 to 10 for temperate crops). Watson (1947) presented values of $\mathbf{L}$ which, although low by present standards, conveniently illustrate the nature of seasonal trends. In *Beta vulgaris* (sugar beet), *Hordeum vulgare* (barley), *Solanum tuberosum* (potato) and *Triticum* sp. (wheat) grown at Rothamsted (see Fig. 2.4), $\mathbf{L}$ was closely related to the time of planting of the crop, and to its subsequent phenology. Unlike $\mathbf{E}_A$, it was largely independent of seasonal changes in the environment. Watson (1971) has reviewed more recent work in this field.

Since different environmental conditions have a marked effect on the growth and development of stands of vegetation, these conditions are naturally reflected in the values of LAI observed. For example, Black (1963) demonstrated that the optimum LAI for dry matter production in *Trifolium subterraneum* (subterranean clover) was strongly influenced by the level of solar radiation received. The higher this level, the higher the LAI at which maximum production could be sustained. In common with other growth analytical quantities, LAI is also strongly affected by temperature and by the water and mineral nutrient regimes of the crop. Watson (1952, 1971) again provides convenient introductions to these topics. An interesting and recent development

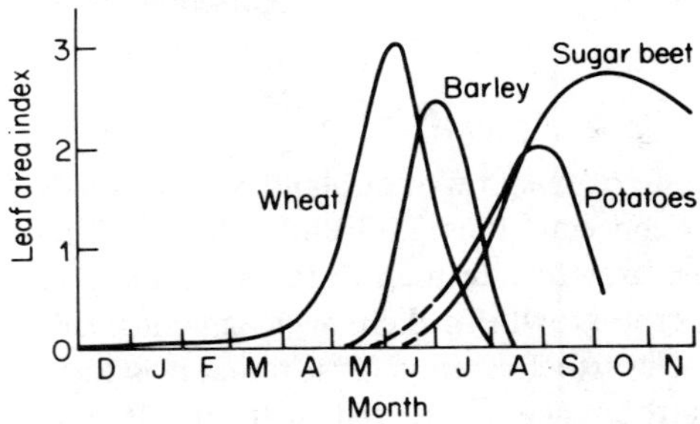

Fig. 2.4 Eye-smoothed seasonal changes in leaf area index in four crops grown outdoors (from Watson, 1947).

is the possibility of determining LAI over large geographical areas from the spectral properties of vegetation photographed from artificial satellites (Wiegand, Richardson and Kanemasu, 1979).

2.3.3 Crop growth rate

Straightforward reasoning suggests that overall yield is controlled both by the efficiency of the leaves of the crop as producers of dry material and by the leafiness of the crop itself. This relationship may be given a precise notation. Since unit leaf rate, **E**, is defined as

$$\mathbf{E} = \frac{1}{L_A} \cdot \frac{dW}{dT} \qquad (2.9)$$

and leaf area index, **L**, is defined as

$$\mathbf{L} = \frac{L_A}{P} \qquad (2.31)$$

we can write

$$\frac{1}{P} \cdot \frac{dW}{dT} = \frac{1}{L_A} \cdot \frac{dW}{dT} \times \frac{L_A}{P} \; . \qquad (2.32)$$

The new quantity on the left-hand side of equation 2.32 is the instantaneous rate of dry matter production per unit area of land, a simple and important index of agricultural productivity or rate of dry matter production. Although various workers had previously made use of this concept, it was first given a name, crop growth rate, by Watson (1958). The contraction CGR is now in general use and **C** is a convenient symbol. Its units are weight land area^{-1} time^{-1}. The equation

$$\mathbf{C} = \mathbf{E} \times \mathbf{L} \qquad (2.33)$$

is the central relationship in the study of population and community growth, just as

$$\mathbf{R} = \mathbf{E} \times \mathbf{F} \tag{2.13}$$

is central to the study of the growth of plants as individuals. One cannot press the similarity between equations 2.33 and 2.13 too far since **C** is closer in concept to absolute growth rate than to **R**, and **L** and **F** may each be applied to the analysis of crop growth in their own right, instead of being analogues as suggested here. Nevertheless, in general terms it is easy to see that in each case the overall performance of the system (**C** and **R** respectively) is broken down into two components: the productive efficiency of its leaves (**E** in both cases) and its leafiness (**L** and **F** respectively).

Perhaps the most satisfactory way of deriving **C** is to multiply together the instantaneous values **E** and **L**, using the relationship defined in equation 2.33. Otherwise, mean crop growth rate, $\overline{\mathbf{C}}$, may be calculated without recourse to **E** and **L** as

$${}_{1\text{–}2}\overline{\mathbf{C}} = \frac{1}{P} \cdot \frac{{}_2W - {}_1W}{{}_2T - {}_1T} \tag{2.34}$$

where ${}_1W$ and ${}_2W$ are the dry weights of crop harvested from equal (but separate) areas of ground, P, at times 1 and 2. If ${}_1W$ and ${}_2W$ are each expressed per unit quantity of P then equation 2.34 can be simplified to

$${}_{1\text{–}2}\overline{\mathbf{C}} = \frac{{}_2W - {}_1W}{{}_2T - {}_1T} \quad . \tag{2.35}$$

Mean crop growth rate in this form indeed becomes an absolute growth rate – a difference in size divided by a difference in time (see section 2.2.2 and equation 2.2). For this reason, despite the presence of two variates in addition to time in the instantaneous definition, CGR is perhaps best thought of as a 'type (i)' rather than a 'type (iii)' quantity. Equation 2.32 then becomes another instance of '(i) = (ii) × (iii)' (see section 2.1).

The calculations for the mean values, $\overline{\mathbf{C}}$, $\overline{\mathbf{E}}$ and $\overline{\mathbf{L}}$ all depend on various unrelated assumptions. Only rarely do these concur, so, in most cases,

$$\overline{\mathbf{C}} \neq \overline{\mathbf{E}} \times \overline{\mathbf{L}} \tag{2.36}$$

for just the same type of reason that

$$\overline{\mathbf{R}} \neq \overline{\mathbf{E}} \times \overline{\mathbf{F}} \ . \tag{2.15}$$

In practice this means that although equation 2.36 cannot properly be used to *calculate* any one quantity from estimates of the other two, the relationship is often close enough to be valuable in *interpreting* the overall growth of the system in terms of its component processes.

As we have seen in section 2.2.4, ULR often declines in magnitude as a crop approaches maturity while LAI (section 2.3.2) normally increases. Indeed, through self-shading the former trend is, at least in part, a consequence of the latter. Now, since CGR is the product of ULR and LAI, the direction and extent of its own drift with time depends on the relative magnitude of these trends. Stoy (1965) grew *Triticum aestivum* (wheat) under controlled environmental conditions and was able to show that as plants grew in the vegetative phase, increasing $\overline{\mathbf{L}}$ caused less than proportionate decreases in $\overline{\mathbf{E}}_A$ (Fig. 2.5) with the result that $\overline{\mathbf{C}}$ increased with time.

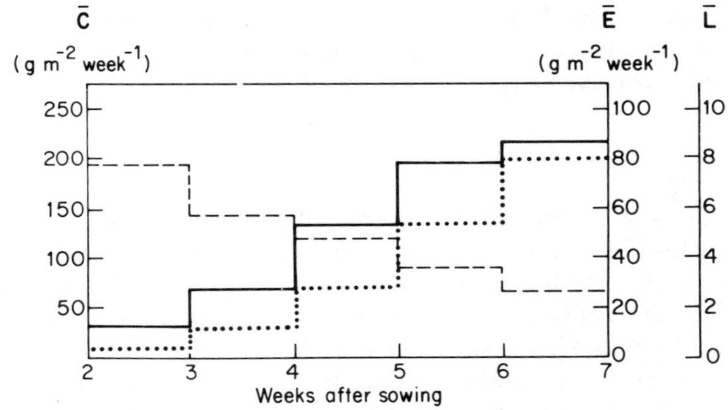

Fig. 2.5 Progressions of mean values of crop growth rate, $\overline{\mathbf{C}}$ (———); unit leaf rate, $\overline{\mathbf{E}}$ (– – – –); and leaf area index, $\overline{\mathbf{L}}$ (· · · ·) in wheat (plotted from data given by Stoy, 1965).

Watson (1952) discussed the relative importance of variation in LAI and ULR in determining CGR and concluded that LAI is on the whole more open to manipulation because it is more strongly dependent on the environmental conditions and management regime of the crop.

Inter-varietal differences are well known, for example Watson (1958), re-working data of Watson (1947), showed that $\overline{\mathbf{E}}$ and $\overline{\mathbf{L}}$ were inversely related in a selection of five varieties of *Solanum tuberosum* (potato). In this particular example, $\overline{\mathbf{E}}$ varied by a greater proportion than $\overline{\mathbf{L}}$ and was thus the more important determinant of $\overline{\mathbf{C}}$.

2.3.4 Leaf area duration and related quantities

When leaf area index is plotted against time (e.g. Fig. 2.4) the resulting curve allows not only an examination of the time-course of this quantity but also an estimate of what Watson (1947) called the 'whole opportunity for assimilation' that the crop possesses during the period in question. Watson suggested that the integral of (the area lying beneath) the LAI versus time curve might conveniently be called leaf area duration, LAD, since it 'takes account both of the magnitude of leaf area and its persistence in time'. In effect, it represents the leafiness of the crop's growing period. (I am grateful to J. Květ for the information that this quantity is termed 'photosynthetic potential' by Soviet growth analysts.)

The single symbol in use for LAD is **D**. Being an integral quantity, LAD cannot be defined instantaneously. Instead, its definition is

$$_{1-2}\mathbf{D} = \int_{{}_1T}^{{}_2T} \mathbf{L}\, \mathrm{d}T \ . \tag{2.37}$$

Two methods of estimating **D** across a period of time are, firstly, the numerical integration of a function describing the progress of **L** versus time (see section 3.4.6) and, secondly, a graphical approach, the simple area

$$_{1-2}\mathbf{D} = \frac{({}_1\mathbf{L} + {}_2\mathbf{L})({}_2T - {}_1T)}{2} \ . \tag{2.38}$$

This most closely represents reality when the second derivative (see section 3.4.7), $\mathrm{d}^2\mathbf{L}/\mathrm{d}T^2$, is near zero, necessitating short harvest intervals where the curvature of **L** versus time is marked. Since the units of any area are the product of the units of each of its two dimensions, and since LAI is a dimensionless ratio, the units of LAD are the same as those of time.

If the LAD of a crop and its mean ULR are known then its final yield may be predicted. Less perversely, if this yield is already known (as it would be if ULR had been derived) then it may be broken down into these two components:

$$\underset{(\text{weight area}^{-1})}{\text{Yield}} \approx \underset{(\text{time})}{\text{LAD}} \times \underset{(\text{weight area}^{-1}\ \text{time}^{-1})}{\text{ULR}} \tag{2.39}$$

The approximation sign is used here because the concept of mean unit leaf rate over a whole season is inevitably very crude and because, in the concept of leaf area duration itself, equal areas beneath the LAI curve are treated as being equally useful 'opportunities for assimilation' – an even more dubious assumption in view of the changes that take place, both ontogenetically within the crop and climatically in its environment, during the course of the crop's growth. Evans (1972, p. 224) discusses these problems further.

Another way to derive LAD is to estimate the area under the curve of L_A (as opposed to LAI) versus time. This produces another LAD, with units of area × time. For example, these might be $\mathrm{m}^2 \cdot$week ('square metre weeks'). In effect, this is a LAD expressed in terms of leaf area per plant, instead of the leaf area per plot ('ground area equivalents') used in the previous section. This change in units is reflected in the general relationship

$$\underset{\text{(weight)}}{\text{Yield}} \approx \underset{\text{(area} \times \text{time)}}{\text{LAD (leaf area basis)}} \times \underset{\text{(weight area}^{-1}\text{ time}^{-1}\text{)}}{\text{ULR}} \, . \tag{2.40}$$

Yield, like LAD, is now expressed on a per crop basis. The same inexactness applies here as applies to equation 2.39. All of these 'durations' are of 'type (iv)' (section 2.1).

Using the familiar four species grown at Rothamsted, Watson (1947) demonstrated that LAD was a more important factor in determining final yield than mean ULR (Table 2.3). Equation 2.40, with yield and LAD as known quantities, was used to derive mean ULR. The analyses for sugar beet run only until the end of October, which is rather short of its whole growing season. The time of year at which most of the crop's foliage is displayed is of some importance. Obviously, other things being equal the greatest possibilities for high production occur when a substantial LAI coincides with the mid-summer conditions, where ULR is maximal (cf. section 2.2.4, and Fig. 2.4). In Watson's group of four species this conjunction was seen only in barley.

Table 2.3 A comparison of yield, leaf area duration and unit leaf rate in four crops grown at Rothamsted. (From Watson, 1947.)

	Yield tonne ha^{-1}	LAD weeks	Mean ULR tonne ha^{-2} week^{-1}
Hordeum vulgare (barley)	7.3	17	0.43
Solanum tuberosum (potato)	7.7	21	0.36
Triticum sp. (wheat)	9.5	25	0.38
Beta vulgaris (sugar beet)	12.0	33	0.36

A number of concepts analogous to LAD may be envisaged. For example, Kvĕt, Svoboda and Fiala (1969) derived a biomass duration, BMD, to which Kvĕt *et al.* (1971) gave the symbol **Z**. This was defined as the area under the weight versus time curve in the same way that LAD was defined as the area under the same curve for LAI or (L_A). The units of BMD are weight × time. This quantity is related to total yield, not by mean ULR, the productive efficiency of unit amounts of leaf area, but by the mean RGR itself, the productive efficiency of unit amounts of dry matter.

$$\underset{\text{(weight)}}{\text{Yield}} \approx \underset{\text{(weight} \times \text{time)}}{\text{BMD}} \times \underset{\text{([weight weight}^{-1}\text{] time}^{-1}\text{)}}{\text{RGR}} \, . \tag{2.41}$$

This relationship is also inexact (cf. equations 2.39 and 2.40) but as a crude summary of the behaviour of the crop it can be useful (Kvĕt and Ondok,

1971). Květ, Svoboda and Fiala (1969) described BMD as 'being an approximate measure of the stand's vitality' and pointed out that

$$\underset{(\text{area} \times \text{time})}{\text{LAD (leaf area basis)}} \approx \underset{(\text{weight} \times \text{time})}{\text{BMD}} \times \underset{(\text{area weight}^{-1})}{\text{LAR}} \; . \qquad (2.42)$$

Parallel developments, using length, volume, total chlorophyll or calorific content as primary data, will be obvious.

2.4 Independent variables or variates other than time

Although these are outside the defined scope of this book, in the interests of completeness of coverage of plant growth analysis it is worth considering, briefly, the interpretation of progressions plotted against quantities other than time.

Where, for any one species, derived quantities are plotted against time (e.g. Fig. 2.4), the effects of any experimental treatments may accelerate or decelerate the natural ontogenetic drift in the quantity under examination. Hence, plants of an age but in different treatments are not necessarily all in the same morphogenetic condition. To overcome this disadvantage (minimizing the effects of ontogenetic drift) it may sometimes be useful to plot the derived quantities against another index of development, such as total dry weight or number of leaves, instead of against time (see, for example, Austin, 1963).

In practice this may be achieved if values of the derived quantities, like **R** and **E**, are plotted in order of, say, increasing W. Naturally, this approach is more easily executed using instantaneous values. Indeed, if W does not figure in the derived quantity itself (e.g. SAR, $(1/R_W)(\mathrm{d}M/\mathrm{d}T)$ it may actually be inserted into the analyses in place of T (giving $(1/R_W)(\mathrm{d}M/\mathrm{d}W)$). However, these arguments are as broad as they are long: what guarantee is there that plants of one dry weight in different treatments have reached that dry weight by similar routes and are currently in the same morphogenetic condition?

There is also another danger here. When quantities derived from W are plotted against W or against other quantities derived from it, misleading relationships can be obtained. If, for example, **R** is roughly constant, plotting **E** against **F** (that is, against **R**/**E**) can produce a straightish line for mathematical reasons alone. This can obscure a quite genuine dependence of **E** on **F** when increases in **F** lead to self-shading, thus causing a fall in **E**.

One noted developmental time-scale, of long standing in the literature, is the plastochron index. A plastochron is the period of time between the initiation of (normally) one leaf and the next in the shoot apex. Plant age on the basis of the plastochron is expressed as the number of such intervals that have elapsed at any given time. Michelini (1958) pointed out with some pungency the disadvantages attached to the use of a time-based abscissa when investigating fresh weight, chlorophyll content and respiratory behaviour in successive individual leaves. He presented convincing progressions on a plastochron time scale (to which he also fitted segmented regressions, see section 7.2). Other work employing this type of approach is that of Bond (1945) and Erickson (1959).

Various attempts have been made, in the relatively standardized world of crop physiology, to construct a developmental scale involving other organs in addition to leaves. Such a scheme was put forward by Zadoks, Chang and Konzak (1974) in the form of a decimal scale of principal and secondary stages of growth, potentially 100 points long. This has been illustrated, with stylized drawings of selected stages in the growth of wheat, barley and oat (and with expanded descriptions of some of the stages), by Tottman, Makepeace and Broad (1979). Numbers of root nodules in legumes have also been explored for use as a developmental scale (e.g. by Blum and Heck, 1980).

In addition to methods which aim to recognize the natural developmental state of the plant there have been many devised which take account of the accumulated experience of the plant in respect of selected environmental variates. Nelder *et al.* (1960) devised a scheme with the general aim of

> 'replacing chronological time in the growth equations of vegetative crops by a time scale based on some suitable combination of meteorological factors.'

Emecz (1962) considered growth in relation to the level of available photosynthetically-active radiation and Nichols (1972) tested the utility of logistic models (section 6.3) fitted on a variety of 'heat unit' time scales (calculated as temperature integrals above a range of base temperatures derived from daily maxima and minima).

Finally, we note that a substantial field of activity exists in which other independent variables or variates are used not instead of, but in addition to, time. These are specially discussed in section 7.7.

2.5 Synopsis and sources of additional information

Table 2.4 contains a list of symbols used for the primary data; Table 2.5 contains a synopsis of all the terms, contractions, symbols, definitions and units relevant to the concepts introduced in this chapter, Table 2.6 contains generic formulae for mean values of these quantities for the harvest interval ${}_1T - {}_2T$. For a guide to the principles followed in allocating symbols, see section 1.8.

In the preceding sections of this chapter I have attempted to provide a conceptual and semi-practical introduction to the derived quantities of plant growth analysis. There are a number of other synopses of this type, varying greatly in length and emphasis, on which the reader may care to draw.

Table 2.4 Symbols used for primary quantities in plant growth analysis (see also Tables 2.5 and 2.6).

Symbol	Quantity
A	area
FW	total fresh weight
L_A	total leaf area
L_W	total leaf dry weight
M	mineral nutrient content (of one particular element or group of elements)
P	ground area per sample
PN	protein nitrogen content
RNA	ribonucleic acid content
R_W	total root dry weight
S_W	total shoot dry weight (above-ground parts of the plant)
T	time
W	total dry weight
w_i	dry weight of the ith component
W_P	dry weight of perennating structure
Y	generalized dependent variate
Z	generalized dependent variate

The first of these was a review article by Watson (1952). This dealt chiefly with the population and community levels of organization and covered unit leaf rate, growth in leaf area and leaf area index in field crops. Blackman's (1961) chapter in M. X. Zarrow's *Growth in Living Systems* gave an historical introduction to the subject, with a review of the effects of specific environmental factors at the organismal level. Radford (1967) provided a short, clear explanation of the assumptions involved in the use of the classical formulae (Table 2.6) and also a brief introduction to the functional approach. Complementing his 1961 publication, Blackman (1968) provided a review chapter covering the process of dry matter production at the population and community levels of organization. Watson (1968), in a presidential address to the Association of Applied Biologists, reviewed the history of, and future for, growth studies on field crops. The process of dry weight increase in individuals was covered from a mathematical point of view by Richards (1969) in a textbook chapter that remains unique in its depth of coverage of this special topic. In 1971 Květ *et al.* provided the longest and most successful review to date, covering both the classical and the functional approaches. The first whole book on the subject, by Evans (1972), remains the definitive work for the classical approach. It also provided a brief introduction to the functional approach and included much coverage of related topics such as experimental design and procedure, environmental measurement and control, and gas-exchange studies.

Aside from the special issues surrounding the analysis of quantal responses (all or none, see Hewlett and Plackett 1979), many authors of textbooks of plant physiology include sections on describing and interpreting the growth of whole organisms, but these are mostly too slight to warrant inclusion here. Exceptions are Leopold and Kriedmann (1975), who devoted a whole chapter (pp. 77–105) of their *Plant Growth and Development* to 'the dynamics of growth'. The result was a perceptive treatment of the main concepts of both individual and population growth analysis, with substantial illustration of environmental influences. As an introductory chapter to *The Shoot Apex and Leaf Growth*, Williams (1975) provided an excellent introduction to relative growth rate and the process of dry weight increase. Causton (1977) included a chapter on plant growth analysis in *A Biologist's Mathematics*; by way of a biological introduction to the use of the calculus he provided a brief but clear coverage of the growth analysis of individuals.

Hunt (1978a) offered an introductory account of all of plant growth analysis in short textbook form. The material presented in section 2.2 and section 2.3 was drawn from the 'classical' parts of this source, the chief difference being that the longer account given by Hunt (1978a) was presented at a slightly 'lower' level, and with rather more illustration. Causton and Venus (1981) included a chapter covering both the classical and the functional approaches, but only for the analysis of the growth of individuals and plant parts. This was comprehensively illustrated with their own experimental data. Their chapter performs much the same role in *The Biometry of Plant Growth* as does the present one here, considerable differences in emphasis notwithstanding. In a symposium chapter, Hunt (1981) covered in 16 pages the ground the present book covers in 248.

For non-English-speaking workers there are reviews of the classical approach in Japanese (Saeki, 1965) and in Czech (Repka and Kostrej, 1968).

Table 2.5 A synopsis of derived quantities involved in plant growth analysis.

Quantity	Contraction	Symbol	Definition	'Type' (see sect. 2.1)	Dimensions, if any	See page
Absolute growth rate	AGR	**G**	$\frac{dW}{dT}$	(i)	$W\,T^{-1}$	16
Biomass duration	BMD	**Z**	$\int_{T_1}^{T_2} W\,dT$	(iv)	$W\,T$	39
Component production rate	CPR	**J**	$\frac{1}{W} \cdot \frac{dw_i}{dT}$	(iii)	$W\,W^{-1}\,T^{-1}$	32
Crop growth rate	CGR	**C**	$\frac{1}{P} \cdot \frac{dW}{dT}$	(i)*	$W\,A^{-1}\,T^{-1}$	35
Fresh weight-dry weight ratio	FWR	–	$\frac{FW}{W}$	(ii)	–	28
Leaf area duration (leaf area basis)	LAD	**D**	$\int_{T_1}^{T_2} L_A\,dT$	(iv)	$A\,T$	38
Leaf area duration (leaf area index basis)	LAD	**D**	$\int_{T_1}^{T_2} \mathbf{L}\,dT$	(iv)	T	37
Leaf area index	LAI	**L**	$\frac{L_A}{P}$	(ii)	–	33
Leaf area ratio	LAR	**F**	$\frac{L_A}{W}$	(ii)	$A\,W^{-1}$	23
Leaf weight ratio	LWR	–	$\frac{L_W}{W}$	(ii)	–	26

Rate of production of *PN* per unit of *RNA*	–	$\mathbf{G}_{PN,\,RNA}$	$\frac{1}{RNA}\cdot\frac{dPN}{dT}$	(iii)	$W\,W^{-1}\,T^{-1}$	31
Relative growth rate	RGR	**R**	$\frac{1}{W}\cdot\frac{dW}{dT}$	(i)	T^{-1}	16
Root-shoot ratio	RSR or R/S	–	$\frac{R_W}{S_W}$	(ii)	–	28
Root-weight ratio (fraction)	RWR(F)	–	$\frac{R_W}{W}$	(ii)	–	28
Shoot-weight ratio (fraction)	SWR(F)	–	$\frac{S_W}{W}$	(ii)	–	28
Specific absorption rate	SAR	**A**	$\frac{1}{R_W}\cdot\frac{dM}{dT}$	(iii)	$W\,W^{-1}\,T^{-1}$	30
Specific leaf area	SLA	–	$\frac{L_A}{L_W}$	(ii)	$A\,W^{-1}$	26
Specific utilization rate	SUR	**U**	$\frac{1}{M}\cdot\frac{dW}{dT}$	(iii)	$W\,W^{-1}\,T^{-1}$	30
Unit leaf rate (= net assimilation rate)	ULR (= NAR)	**E**	$\frac{1}{L_A}\cdot\frac{dW}{dT}$†	(iii)	$W\,A^{-1}\,T^{-1}$	23
Unit production rate	UPR	Π	$\frac{1}{W_P}\cdot\frac{dW}{dT}$	(iii)	$[W\,W^{-1}]\,T^{-1}$	31
Unit shoot rate	USR	**B**	$\frac{1}{S_W}\cdot\frac{dW}{dT}$	(iii)	$[W\,W^{-1}]\,T^{-1}$	29

See Table 2.4 for definitions of symbols used for primary quantities and Table 2.6 for formulae that can be used to derive mean values over the interval ${}_1T$ to ${}_2T$.

*Not 'type (iii)', see p. 36.

†Many variates other than L_A may be used, see p. 25.

Table 2.6 General formulae for mean values of the quantities listed in Table 2.5, each estimated over the time interval ${}_1T$ to ${}_2T$. Except in the case of type (i) the formulae are approximate.

Type of quantity	Generalized definition	Formula for mean value
(i)	$\dfrac{\mathrm{d}Y}{\mathrm{d}T}$	$\dfrac{{}_2Y - {}_1Y^*}{{}_2T - {}_1T}$
(i)	$\dfrac{1}{Y} \cdot \dfrac{\mathrm{d}Y}{\mathrm{d}T}$	$\dfrac{\log_e {}_2Y - \log_e {}_1Y}{{}_2T - {}_1T}$
(ii)	$\dfrac{Y}{Z}$	$\dfrac{({}_1Y/{}_1Z) + ({}_2Y/{}_2Z)}{2}$
(iii)	$\dfrac{1}{Z} \cdot \dfrac{\mathrm{d}Y}{\mathrm{d}T}$	$\dfrac{{}_2Y - {}_1Y}{{}_2T - {}_1T} \cdot \dfrac{\log_e {}_2Z - \log_e {}_1Z}{{}_2Z - {}_1Z}$
(iv)	$\int_{{}_1T}^{{}_2T} Y \, \mathrm{d}T$	$\dfrac{({}_1Y + {}_2Y)({}_2T - {}_1T)}{2}$

*for crop growth rate, set $Y = W/P$

3

The functional approach in theory

3.1 Introduction

In the preceding chapters I have tried to explain how and when plant growth analysis may be a useful experimental tool and have given an outline of its main concepts. Now it is time to turn to the central topic of this book, the derivation of the various growth-analytical quantities from mathematical functions fitted to the primary data. It is assumed that by this stage the experimenter has examined his own needs sufficiently thoroughly to decide whether or not the general idea of a 'growth analysis' is relevant to the questions he is asking. If this decision is 'yes', then this chapter is the starting point of the experimenter's functional approach. This chapter will examine what the functional approach is and what it is not and will place the approach into the context of other modelling activities in plant science. It will describe, mainly in general terms, the ways in which estimates of the quantities catalogued in Chapter 2 may be obtained from fitted growth functions and it will attempt to take stock of the value of the data in their new, processed form. Some of this ground was covered by Hunt (1979).

3.2 Models of plant growth

The functional approach to plant growth analysis is a branch of mathematical modelling. Now, a model is simply something constructed like something else and a mathematical model is a model constructed of mathematics. So, a mathematical expression or group of expressions that behaves in some way like a real system can be called a mathematical model of that system.

Thornley (1976) has written at some length about mathematical models in plant physiology and emphasized that such models fall into two broad divisions. Firstly, there are mechanistic models. These are conceived in terms of the mechanism of the system, or how the parts of the system work together as they might in a machine. An extremely simple example of a mechanistic model is the relationship

$$C = f(D)\,. \qquad (3.1)$$

Here, the circumference of a circle, C, is expressed as a function of its diameter,

D. It has been known since ancient times that the 'function' in question is merely a constant multiplier. This constant has been called π and its value has been found, to four significant figures, to be 3.142. Despite being so trivially simple, the model is sound. In it, *C* and *D* are forever locked together in an unvarying relationship, like cogs in a machine. One expects, and can show for circles of all sizes, that the model is an exact representation of reality.

A second type of mathematical model is the empirical model. Here, the mathematics simple re-describes the data from which the model was constructed without giving rise to any information that was not contained in the original data themselves. One example of an empirical model, again absurdly simple, is an expression relating the annual number of deaths, dN_m/dT, to total population size, N, in the human population of the United Kingdom in 1971:

$$\frac{dN_m}{dT} = 0.0119\,N \ . \tag{3.2}$$

This model merely re-describes a vast set of census data in a condensed mathematical form. It states that the annual rate of mortality, $(1/N)(dN_m/dT)$, is, in the demographers' terminology, 11.9 individuals per thousand. Why this rate should have this value and not some other is evident neither from the original data nor certainly from the model. The rate depends on a bewildering set of social, economic and medical factors, operating both now and in the past. It is simply the value returned by the appropriate processing of the primary data. Of interest, however, would be a comparison of similarly-treated data between different sampling occasions or between different nations at the same sampling time. For example, Ehrlich and Ehrlich (1972) contrast the figure given above with one of 12.4 deaths per thousand per annum in Belgium and 5.0 in Hong Kong, holding these differences to be due to the slightly more old-biased and substantially more young-biased age distributions of these two populations respectively, in comparison with that of the United Kingdom. This important role of the empirical model as a comparative tool (Heath, 1932; Thornley, 1976) will be returned to.

But first we must consider carefully the other differences between mechanistic and empirical models. Thornley (1976) has concluded that, in the whole field of plant physiology,

> 'It needs to be stressed that there is no clearly defined dividing line between the two methods, and it is usual for most modelling exercises to contain both empiricism and mechanism in varying admixtures. It is more a matter of emphasis. The mechanistic modeller will tend to construct his models before doing the experiments, thinking of possible mechanisms and deducing their consequences by means of a model; the experiment will then test his hypotheses, and possibly favour one mechanism rather than another.

However, in thinking of mechanisms *a priori*, he is of course guided by existing data and knowledge, and to these he applies his own blend of empiricism and intuition. On the other hand, the empirical modeller may well make his guesses about mechanism after doing the experiment and looking at the data, so he begins an investigation as an empiricist and ends up as a mechanist. In practice the modeller swings like a pendulum between mechanism and empiricism; he tries to make progress whenever and however he can, although he should prefer mechanism to empiricism as a general rule, and if the problem allows it.'

In the more specific context of the modelling of plant growth, we must recognize that a highly complex web of processes comprises the whole system. The great difficulty of representing these processes as a simple mechanistic model was squarely faced by Richards (1969) in a review of the earlier attempts to analyse plant growth from a mechanistic point of view. For example, the monomolecular function had been used to predict W, the dry weight of the whole plant at any time T, from a knowledge of a final dry weight, a, to which the plant is ultimately limited:

$$W = a(1 - be^{-cT}) \quad . \tag{3.3}$$

The absolute growth rate, dW/dT, is $c(a - W)$. This is proportional to the amount of growth *yet to be made* and declines linearly in value with increases in W. This decline proceeds at a rate determined by the coefficient c, which is the inherent capacity of the plant to increase its dry material. The coefficient b is a measure of the material present at the beginning of growth, such that when $T = W = 0$, then b has a value of 1. Not the least of the difficulties involved in the use of this formula is the concept of a final, limiting value. Causton (1977) has pointed out that much of plant growth is indeterminate since the ability of a plant to produce new material is not usually circumscribed in the same way as is, for example, the production of both the parts and the whole of most animals, or the progression of a chemical reaction which halts when all of the reagents have been consumed (the equation was originally introduced into biology from physical chemistry). While acknowledging that, from time to time, such simple mechanistic models 'are capable of reproducing the course of growth curves with tolerable accuracy, and sometimes very closely' Richards (1969) concluded that in the search for realistic equations of growth 'the outcome has been disappointing.' Williams (1964) stated that 'it has never been possible to show that any one of them fits the facts so exactly that the others can be excluded' and the same author, with D. Bouma in 1970, suggested that 'the fitting of continuous functions to extensive growth data [in an attempt to achieve a mechanistic model of growth] is rarely justified, for it is unreasonable to expect growth to be governed by a single set of parameters throughout its course, even in a controlled environment.' Unreasonable indeed, and one is led to the conclusions that whole plants rarely behave as simple machines (except perhaps over short periods of time) and that simple, comprehensive mechanistic models of the type attempted in equation 3.3 do not exist. Considerably greater elaboration of the models is needed for them to approach reality with any conviction. The chances of the models doing so are further increased if the scope of the exercise is confined to certain sets of conditions, or to the growth of less than the whole plant. Recent developments in mechanistic growth-modelling which should be taken into account by readers keen to continue in this direction are the models for the growth in water culture of *Lycopersicon esculentum* (tomato) by Thornley and Hurd (1974), and of *Allium cepa* (onion) by Nye, Brewster and Bhat (1975); for plant growth as a general process, including senescence, by Thornley (1976 p. 196); for crop growth as a function of time and planting density by Barnes (1977); for growth and partitioning of assimilates in grasses by Troughton (1977); for cotton yield by Wallach (1978, 1980); for forage production and digestibility by Edelsten and Corrall (1979) and for crop growth as a general process by Charles-Edwards and Fisher (1981). Methodological guidelines in this field have been given by Calow (1976), Thornley (1976), Jeffers (1980) and Rose and Charles-Edwards (1981).

In the fields of animal and human biology the modelling of growth on a semi-mechanistic basis has also been extensively attempted (Clark and Medawar, 1945). The discussion led by Zuckerman (1950) presents many viewpoints, including a perspective by F. G. Gregory on 'Growth and form in plants'. The whole discussion amounts to a most useful synthesis of the theoretical and practical foundations of what might analogously be called 'animal growth analysis'.

In this brief overview of models in plant growth it will be evident that the word 'model' has deliberately been used in one of its wider senses. There are those who would not admit the empirical equation to this title, those who would admit both empirical and mechanistic representations but disregard the empirical as valueless, those who would admit both and try to respect each for its own particular purpose, and those so disenchanted with models attempting to be mechanistic that whole-system empirical equations remain the limit of their endeavours. The reader new to the field can do no better than to read more from the first chapter of Thornley (1976) for a balanced appreciation of the broad differences between the two types of model.

Lest the foregoing should have given the impression that mechanistic models are often only well-intentioned failures and that empirical models are often only pragmatic working arrangements it should be remembered that under the best of circumstances even failed mechanistic models may draw attention to those parts of the system which are least well understood and that the most mundane empirical model may occasionally uncover an important and fundamental relationship if, for example, the primary data were too variable to suggest the same without treatment. But since it is to empirical models that most of the remainder of this book is devoted, it is of the greatest importance to understand that even when the reductionistic pursuit of mechanistic insight into plant growth is consciously abandoned by the modeller, that which remains is substantial. I am aware that this statement directly conflicts with the viewpoint, which is a common one, of Waddington (1956, as quoted by Williams, 1964):

> 'If we can use these formulae "merely as convenient means of summarizing the empirical observations", the theoretical gain has not been great.'

I hope to show during the course of this text that in many ways this is untrue, and that empirical models of growth can benefit experimental plant science in a unique way.

3.3 Models in the functional approach to plant growth analysis

In moving from mechanistic to empirical models of plant growth

> 'the particular mathematical form of the function used is now regarded as of no special physiological significance, but accuracy in the fit achieved becomes the primary aim. The resulting equation for the curve summarizes the growth data in a convenient way; moreover, the original data, disturbed by irregular errors, are replaced by a smooth continuous function' (Richards, 1969).

When pursuing a plant growth analysis these mathematical functions are used to provide *fitted* values of data which may in turn be used to derive instantaneous values of the various growth analysis quantities described in Chapter 2. I shall return to the ways in which these quantities can be derived from fitted functions later in this chapter but, firstly, a general philosophy underlying the use of fitted functions will be put forward.

The purpose of an empirical model is to describe reality in a useful way. When making this statement, the view of reality that I adopt is that of the working scientist: reality is the complex abstraction that forms in the mind of the experimenter from the integration both of measurement and of sensory stimulus. What may lie behind this abstraction is, fortunately, not important to us because, in the words of Wigglesworth (1967),

> 'science is concerned with "verifiability"; it is not concerned with ultimate "truth".'

In the physical sciences what constitutes this kind of reality is seldom in doubt. Measurement here is for the most part accurate and easy. Naturally, those at the frontiers of these subjects would disagree with this assertion, but it cannot be disputed that in everyday experience the biologist confronted with the need to make physical measurements finds fewer problems than when confronted with the need to make biological ones. When Evans (1976,

p. 6) referred to 'the inaccessible plant' he emphasized that, particularly in its natural environment, to approach the plant for the purpose of observing and quantifying its real and natural performance requires the greatest care if the experimenter is not to

> 'interfere destructively with the subject of the investigation, altering either the environment, or the plant, or both.'

Even 'non-destructive' measurements can be deleterious to subsequent growth (Beardsell, 1977).

This difficulty means that in the physical and the biological sciences, observational data have a rather different standing. In both, their objective is to describe reality, which can be assumed to be incapable of completely perfect description, existing only as descriptions of varying states of imperfection. But, whereas most physical data may usually be acceptable as a close enough estimate of this reality, the difficulty in making true and accurate observation in biology, and the inherently more variable state of the subject material, makes this process of estimation much less satisfactory. Here is the crux of the problem and the greatest single opportunity for the functional approach to plant growth analysis to make its contribution.

Suppose that a mono-specific population of plants is being grown for the purpose of defining the growth rate of the species. This might be to compare it with some other species or subspecies, or to investigate the effects of the environment, or whatever; the experimenter at any rate needs a single value of, say, absolute growth rate per individual, dW/dT, which will characterize the species' performance in the trial. This is not a vain endeavour, such a rate truly exists for the population; it is a real and unique parameter of its growth. The question is: How best to estimate it?

Since one is seeking a rate of growth then measurements of weight on at least two occasions are needed. But the taking of the dry weight of a plant is a destructive process and so the same plants cannot be used for the second harvest as were used for the first. Some kind of subsampling is required. The simplest design would be to take half of the population for the initial estimate of weight and the remaining half for the second. But the first subpopulation would not necessarily be representative of the whole population at the time of the first harvest, and the second subpopulation would not necessarily be representative of the state of the original population had the whole of it survived to the end. This is already a source of error. When one considers that in the experimental environment, in the layout of plants in the experiment and in the harvesting techniques employed there are likely to be both accidental and unavoidable differences between the treatment of different individuals within each subpopulation, and when one also considers that even in material derived from a clonal source, different individuals will grow at slightly different rates (Burdon and Harper, 1980), then the sum of these sources of error will not be negligible. Under these circumstances the observational data can only be an approach to reality, sometimes poor, sometimes good, but never the *definition* of reality that some experimenters may misguidedly expect. Anyone who doubts that this is true should try repeating the experiment as exactly as possible; he would end up with two 'realities'. Fisher (1966, 1970) and Heath (1970) have provided accounts of this variability and its origins.

How can the functional approach to plant growth analysis ease the difficulty? The rationale is simple: if attempts to assess the reality of growth result in a random scatter of observations about that reality, then a mathematical function fitted to those observations may be expected to regain much of the clarity with which the reality is perceived by the experimenter. In a sense, the course of the flow of understanding is reversed and the fitted function reflects back – not perfectly of course, but at least in the right direction – towards that reality of which the observational data are an imperfect estimate. Paradoxically, the fitted function can be of more value to the experimenter than the data from which it was derived. If this book has a centre, then it is here.

Naturally, if the fitting of the function is handled clumsily, then the experimenter's perception of reality will flow still further along its corrupted course. This is why it is vital for the fitting of growth functions to be restricted to data believed to be subject only to random errors. If the errors are systematic then a serious distortion of the truth is introduced. Perhaps this is why some experimenters mistrust data 'processed' in this way and attach a reverential affection to 'plain' data, however subject to experimental error. I hope to show during the course of my argument the wrongness of this viewpoint.

To return to the specific, if hypothetical, example given above, one could expect the slope of the regression of dry weight on time to yield the best estimate of absolute growth rate per individual. In symbols, if

$$W = f(T) \tag{3.4}$$

then

$$\frac{dW}{dT} = f'(T) \; . \tag{3.5}$$

This example has been constructed because it is the simplest conceivable use of the functional approach to plant growth analysis: no function other than a linear regression may be used because only a single, overall value of dW/dT is needed. In practice, these data may not often justify such a treatment but that is a separate technicality that does not undermine the basic rationale. It can be dealt with later (section 4.7).

Following upon this statement of the first principle of the functional approach to plant growth analysis, I list the subsidiary advantages. Any one of these may alone be a decisive feature, according to the needs of the experimenter.

(1) The model provides a convenient summary of a process which is too complex to understand, or does not need to be understood, in detail but which is of practical significance as a whole.
(2) Much may be said with great economy of expression; a very large body of observational data may be condensed into estimates of a few parameters.
(3) Comparisons between bodies of data different in origin, but similarly treated, are made more straightforward.
(4) Many of the assumptions involved in the calculation of mean values of quantities such as unit leaf rate and leaf area ratio (section 2.2.4) are evaded, the only necessary assumption being that the fitted growth functions adequately describe the primary data.
(5) Information from all sampling occasions is used in determining each value of the derived quantities, whereas the classical method uses data from only two harvests.
(6) The difficulties involved in the pairing of plants (Hunt, 1978a, p. 12) prior to applying the classical formulae are avoided (but see Venus and Causton, 1979a).
(7) The procedure does not depend upon large harvests and the amount of information at risk at each harvest is minimal.
(8) Provided that plants in different experimental treatments are grown simultaneously, the harvesting of these treatments need not be synchronous: interpolated comparisons are feasible.
(9) Replication at different harvests on the same growth curve need not be equal.
(10) Small deviations from the overall trend may be smoothed to gain an impression of growth which is free from random fluctuations (fluctuations believed not to be random in origin may be given separate treatment, see Hunt and Evans, 1980).
(11) Statistical analyses may be integrated into the same analytical procedure as the calculation of the derived quantities.
(12) The approach provides a clearer perception of ontogenetic drift.

Other accounts of the advantages of the functional approach have been given by Vernon and Allison (1963), Hammerton and Stone (1966), Hughes and Freeman (1967), Radford (1967), Ondok and Květ (1971), Evans (1972), Hunt (1973), Hunt and Parsons (1974), Hunt (1978a) and Hunt (1979).

3.4 Derivations from growth functions

3.4.1 Generalized relationships

In section 2.1 we saw how all of the quantities involved in plant growth analysis could be divided into one of four general forms. In plain language these were (i) the absolute or relative rate of production of something, (ii) the ratio of something to something else, (iii) the rate of production of something per unit of something else and (iv) the accumulated duration of something in time.

What follows is an explanation of the ways in which all of these quantities may be derived from growth functions fitted to suitable observational data. For convenience, the particular examples of absolute and relative growth rate, leaf area ratio and unit leaf rate have been chosen when dealing with the growth of individual plants; crop growth rate, leaf area index, unit leaf rate and biomass duration have been chosen when dealing with the growth of populations or communities. The understanding should be that these quantities can take other identities if the analysis is carried out using other primary data. For each analysis I have also given, for completeness here, an alternative which uses a logarithmic transformation of the primary data. The relevance of this transformation, and the situations in which it is likely to be needed, will be dealt with later (section 4.7.2). Moreover, the identity of the fitted mathematical functions used in these analyses is at this stage unimportant; we merely assume that they are sufficient representations of the primary data. A similar, but less extensive, general scheme has been given by Ondok and Květ (1971).

3.4.2 Growth of individuals – untransformed approach

Suppose that whole plant dry weight, W, and total leaf area, L_A, have been measured at varying times, T. Two empirical functions may be constructed:

$$W = f_W(T) \tag{3.6}$$

and

$$L_A = f_L(T) \ . \tag{3.7}$$

then the absolute growth rate is the first derivative or slope of the function describing W (equation 3.6):

$$\frac{dW}{dT} = f'_W(T) \tag{3.8}$$

and the relative growth rate is this quantity divided by W

$$\frac{1}{W} \cdot \frac{dW}{dT} = \frac{f'_W(T)}{f_W(T)} \ . \tag{3.9}$$

Leaf area ratio is simply current L_A divided by current W:

$$\frac{L_A}{W} = \frac{f_L(T)}{f_W(T)} \tag{3.10}$$

and unit leaf rate is equation 3.8 divided by equation 3.7

$$\frac{1}{L_A} \cdot \frac{dW}{dT} = \frac{f'_W(T)}{f_L(T)} \quad . \tag{3.11}$$

3.4.3 Growth of individuals – transformed approach

The primary data L_A and W are transformed to natural logarithms before fitting the empirical functions

$$\log_e W = f_W(T) \tag{3.12}$$

and

$$\log_e L_A = f_L(T) \quad . \tag{3.13}$$

Here, and in the previous section, a function of the form $L_A/W = f(T)$ might be employed instead of one of the above (Hunt, 1973) but in practice there is often no gain, and some difficulty, in doing this (Hughes and Freeman, 1967). To avoid carrying '$\log_e$' through the whole analysis we may re-write equations 3.12 and 3.13 as

$$W = \exp[f_W(T)] \tag{3.14}$$

and

$$L_A = \exp[f_L(T)] \; . \tag{3.15}$$

In contrast to the previous analysis (section 3.4.2) the derivation of relative growth rate is more direct than that of absolute growth rate; RGR is simply the slope of the logarithmic plot of W against time:

$$\frac{1}{W} \cdot \frac{dW}{dT} = f'_W(T) \tag{3.16}$$

and, bringing in equation 3.14 to obtain absolute growth rate,

$$\frac{dW}{dT} = f'_W(T) \cdot \exp[f_W(T)] \; . \tag{3.17}$$

In the case of leaf area ratio, the ratio between the two primary data becomes the difference between the logarithms of these data, hence

$$\frac{L_A}{W} = \exp[f_L(T) - f_W(T)] \tag{3.18}$$

and, employing equation 2.14, unit leaf rate emerges as

$$\frac{1}{L_A} \cdot \frac{dW}{dT} = \frac{f'_W(T)}{\exp[f_L(T) - f_W(T)]} \tag{3.19}$$

which is probably more simply written as

$$\frac{1}{L_A} \cdot \frac{dW}{dT} = f'_W(T) \cdot \exp[f_W(T) - f_L(T)] \tag{3.20}$$

a result which can also be achieved by dividing equation 3.17 by equation 3.15.

3.4.4 Growth of populations – untransformed approach

The experimenter normally has available estimates not of size *per plant,* W and L_A, but of size *per plot,* W/P and L_A/P, expressed per subpopulation of the crop from a known area of ground, P. If our original W and L_A are available, then the primary data WN/P and L_AN/P can be used, where N is the measured density of plants per unit ground area. Assuming the former, we have these two empirical models:

$$\frac{W}{P} = f_W(T) \tag{3.21}$$

and

$$\frac{L_A}{P} = f_L(T) \ . \tag{3.22}$$

These can be written more conveniently as·

$$W = P \cdot f_W(T) \tag{3.23}$$

and

$$L_A = P \cdot f_L(T) \tag{3.24}$$

where P may or may not be unity.

From these, the absolute growth rate per subpopulation is obtained as

$$\frac{\mathrm{d}W}{\mathrm{d}T} = P \cdot f'_W(T) \tag{3.25}$$

and the absolute growth rate per unit of ground area, that is crop growth rate, is

$$\frac{1}{P} \cdot \frac{\mathrm{d}W}{\mathrm{d}T} = f'_W(T) \ . \tag{3.26}$$

We already have an expression for leaf area index

$$\frac{L_A}{P} = f_L(T) \tag{3.22}$$

thus unit leaf rate is simply

$$\frac{1}{L_A} \cdot \frac{\mathrm{d}W}{\mathrm{d}T} = \frac{f'_W(T)}{f_L(T)} \ . \tag{3.27}$$

3.4.5 Growth of populations – transformed approach

Instead of equations 3.23 and 3.24 we take the estimating functions as

$$W = P \cdot \exp[f_W(T)] \tag{3.28}$$

and

$$L_A = P \cdot \exp[f_L(T)] \; . \tag{3.29}$$

Absolute growth rate per plant is then given by

$$\frac{dW}{dT} = P \cdot f'_W(T) \cdot \exp[f_W(T)] \tag{3.30}$$

and crop growth rate itself is

$$\frac{1}{P} \cdot \frac{dW}{dT} = f'_W(T) \cdot \exp[f_W(T)] \; . \tag{3.31}$$

Leaf area index is simply

$$\frac{L_A}{P} = \exp[f_L(T)] \tag{3.32}$$

and unit leaf rate is given by

$$\frac{1}{L_A} \cdot \frac{dW}{dT} = f'_W(T) \cdot \exp[f_W(T) - f_L(T)] \; . \tag{3.33}$$

3.4.6 Durations

The previous four subsections have involved the differential calculus. That is, they have been concerned to a large extent with the slopes of plots of observational data. When dealing with 'type (iv)' durations, however, the integral calculus is needed if the functional approach is to be employed. In practice, this process is less straightforward than the foregoing and, since the classical approach can be made to yield reliable estimates if the data are suitably arranged, experimenters may consider that the functional approach is of less advantage here than elsewhere. However, for those keen to try it the following scheme, using W, is offered.

Biomass duration, $\mathbf{Z}$, is the area beneath the plot of W versus T. To obtain this it is necessary to define two limits, say ${}_1T$ and ${}_2T$ (section 2.3.4). A function describing the basic relationship may be obtained empirically:

$$W = f_W(T) \tag{3.6}$$

and biomass duration be obtained as the definite integral of this function between the limits ${}_1T$ and ${}_2T$:

$$\mathbf{Z} = \int_{{}_1T}^{{}_2T} f_W(T)\,dT \; . \tag{3.34}$$

A difficulty arises if the function itself is not capable of analytical integration. In such cases Causton (1977, p. 152) describes two numerical approaches: (i) the counting of squares beneath a plot of the function – elementary but laborious, and (ii) the application of Simpson's Rule. This involves dividing the time interval ${}_1T$ to ${}_2T$ into an even number of intervals, the more the better, each of width ΔT. A formula is then employed which uses ΔT and the values of the ordinates which separate the intervals. Simpson's Rule offers not only a manual solution to the problem: it can be programmed into a computerized scheme of analysis in which ΔT can be made very small.

No direct method exists for determining BMD from a function describing transformed data:

$$\log_e W = f_W(T) \tag{3.12}$$

since unit areas beneath a logarithmic plot do not represent equal values of weight × time. Instead, such an $f_W(T)$ must be converted mathematically into a function of W (not of $\log_e W$) before any integration can be performed.

3.4.7 Second derivatives

While dealing with the subject of derivations from fitted growth functions, and while suitable functions are close at hand, it is worth mentioning the idea of a second derivative of the original function. Just as a first derivative, such as absolute growth rate, is the equation of the slope of the plot of W against T, so a second derivative of this original plot can be obtained which provides the equation of the slope of the plot of $\mathrm{d}W/\mathrm{d}T$ against T. We no longer examine the rate of change of W, we examine the rate of change *of the rate of change* of W, a quantity which in the conventional notation may be represented as $\mathrm{d}^2 W/\mathrm{d}T^2$. Drawing upon equations 3.6 and 3.9 we have

$$\frac{\mathrm{d}^2 W}{\mathrm{d}T^2} = f''_W(T) \tag{3.35}$$

and, using their logarithmic counterparts, equations 3.12 and 3.16, we can derive the slope of the plot of relative growth rate against time as

$$\frac{\mathrm{d}^2 (\log_e W)}{\mathrm{d}T^2} = f''_W(\mathrm{T}) \ . \tag{3.36}$$

In these two latest equations, of course, the two f_W functions differ in origin.

Second derivatives in the functional approach to plant growth analysis offer a highly condensed summary of the primary data and, although their use may be limited, there will be occasions where they are helpful. For example, Heath (1937a) fitted a quadratic polynomial to the logarithms of W in *Gossypium arboreum* (cotton) growth at Barberton, South Africa:

$$\log_e W = a + b_1 T - b_2 T^2 \tag{3.37}$$

where a, b_1 and b_2 are coefficients. Heath compared the values of the coefficient b_2 for sets of data obtained from two different seasons of growth. In doing so he virtually employed the second derivative of equation 3.37, which is $2b_2$. He was able to establish that a decline in RGR during the 1933–34 season was substantially steeper than a decline in RGR in the 1934–35 season.

Another example of the use of second derivatives is to be found in the work of Idris and Milthorpe (1966), and particularly promising applications have been found in connection with the use of splined regressions (section 7.4).

3.5 The way ahead

With this first departure from generality, and the glimpse it provides of a particular functional model, it is best to close this discussion of the theoretical aspects of the functional approach to plant growth analysis and proceed to the practicalities of curve fitting. From time to time, and where the subject matter warrants it, I shall return to the underlying principles of the approach but for the moment, examples are called for.

4

The functional approach in practice

4.1 Introduction – and a caution

This chapter is mainly concerned with the practicalities of curve fitting. To arrange this material into a logical sequence I have separated curve fitting subjectively by eye from curve fitting by objective, statistical methods and I have also distinguished between curve fitting for the purpose of obtaining values of primary data for insertion into the classical formulae from the activity which forms the bulk of this book, the derivation of instantaneous values of growth-analytical quantities from the fitted curves. To begin with I have also, for completeness, considered the role of curve fitting in the indirect estimation of the primary values needed for plant growth analysis, even though this is technically outside the scope of this work in that it involves independent variables other than time (p. 3). But first, before going to any of this trouble, the experimenter should ask himself a question: Is curve fitting of any sort necessary?

> 'There is a fascination about curve fitting to growth and population data which can all too easily become an addiction' (Williams, 1975).

While I have expounded the advantages of curve fitting in section 3.3, and emphasized that these exist even in the absence of mechanistic motives on the part of the experimenter, I gladly defer to the view that there are situations in which curve fitting may bring no benefit whatever. The experimenter's requirements are either satisfied by the list of advantages given in section 3.3 or they are not. With this caution in mind, examples such as that given (anonymously) in Fig. 4.1, which are not uncommon, deserve a passing mention. Here, the curve fitting was for no particular or implied gain in understanding of the real events that occurred during the experiment other than perhaps to guide the eye through the plots of the primary data. In the instance given, even the data set itself raised doubts about the validity of a regression approach for any purpose at all. Such examples of needless analysis can be avoided if the experimenter is careful to examine his motives beforehand. If none of the advantages given in section 3.3 is sought then curve fitting is unnecessary.

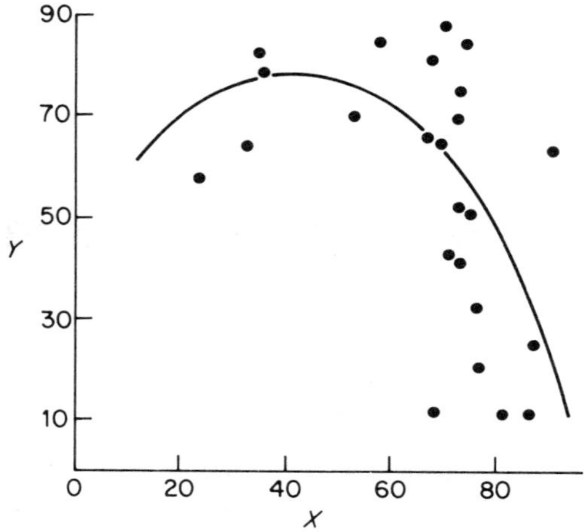

Fig. 4.1 A doubtful regression. Two measurements were linked by the equation $Y = 32.2 + 2.2X - 0.025X^2$ (data and analysis are genuine, published, and reproduced anonymously with permission).

4.2 Indirect estimation of primary data

Since the estimation of many of the quantities needed as primary data in plant growth analysis (for example, dry weights and nutrient contents) involve the destruction of the plant, and since some other estimates (for example, leaf area) are, if not destructive, then difficult to perform without damaging the plant, it is occasionally of great importance to find a way of obtaining these values by some indirect, non-destructive technique. This technique may then be used repeatedly on the same plants or, even if the plants are destroyed, the technique may be substantially less laborious than a direct estimation.

In other branches of science, and in industry, regression equations are in daily use predicting values of some important but inaccessible quantity from observed values of some other quantity which, although of little interest in itself, is of utility as a predictor. The details of these activities need not concern us here since they are adequately covered in applied statistical texts such as Draper and Smith (1966). One example of the use of predicting functions that lies closer to hand is given in the *Forest Mensuration Handbook* of Hamilton (1975). From a knowledge of relatively unimportant but easily obtained quantities like height, diameter at breast height, mid-sectional diameter and top diameter, the forester may estimate more valuable quantities such as timber volume and fresh weight. Provided that values of the estimating variates are obtained in a standardized way, this approach is both rapid and accurate enough for the purpose. The empirical functions are, needless to say, not themselves of concern to the forester in the field: the job is done from tables of solutions of these functions.

In plant growth studies one of the best-documented attempts at indirect estimation of primary data was made by Goodall (1945). Regressions of the dry weight of various components of *Lycopersicon esculentum* (tomato) on the lengths of the leaves and stem were used to follow the changes in size of these components without destroying the plant. Figure 4.2 gives an example of one of these regressions. Similar methods have been used by Goodall (1949), Thorne (1960) and Taylor (1972).

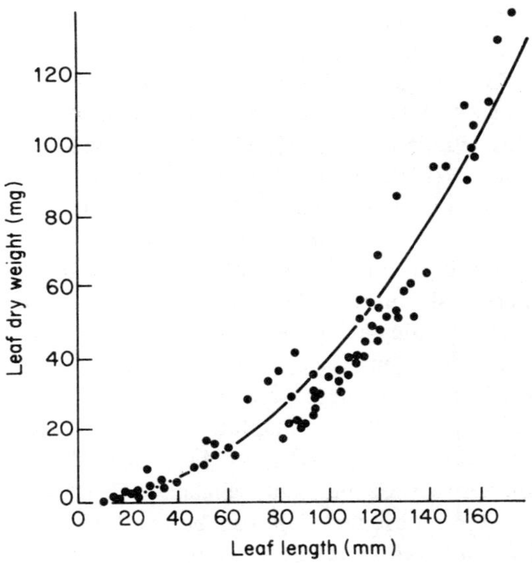

Fig. 4.2 The destructive measurement, leaf dry weight, predicted from the non-destructive leaf length. Data were pooled from a batch of tomato plants, irrespective of leaf age. The fitted regression line is of the form $L_W = bL_L^2$ (parameter value unspecified; data and analysis from Goodall, 1945).

A more recent and comprehensive account of indirect estimation has been given by Ondok (1971). This review is well organized and the section dealing with the use of regression equations includes a table of comparisons of the different regression models that might be considered for use. Among the examples discussed are the calculation of leaf and stem dry weight from measurements of shoot height, stem length and diameter and number of leaves per shoot; the calculation of total crop biomass from number of stems per square metre and stem length; and the total dry weight of individual grass plants from the lengths of the longest green leaves and number of leaves per shoot. In each case an adequate subsample of the population under investigation was used to construct an empirical equation, which was later used to predict values of the quantities of interest to the main experiment. Other workers employing this approach have included Nečas, Zrůst and Partyková (1967), Zrůst, Partyková and Nečas (1974), Hutchings (1975) and Sivakumar (1978).

4.3 Freehand curves for interpolation

4.3.1 In general

The smooth freehand curve enjoys many of the advantages of the functional approach to plant growth analysis not the least of which is the complete independence of any form of computing support, which in some circumstances may be a decisive factor. The other advantages are, using the numeration of section 3.3 (p. 54): (3) 'Straightforward comparisons'; (4) 'Avoidance of assumptions'; (5) 'Whole trends used'; (6) 'Pairings avoided'; (7) 'Frequent small harvests possible'; (8) 'Interpolations possible'; (9) 'Replication may be unequal'; (10) 'Deviations smoothed'; (12) 'Ontogenetic drift made clear'. For those inclined to mistrust empirical mathematical models the freehand curve offers no temptation to read biological significance into the parameters of the model. There is also no need to agonize over the choice of empirical model. On the debit side, the curves are by their nature wholly subjective and a bias can easily creep in unconsciously. Even if this is not a genuine problem, there is no way of proving that this is the case.

In practice the technique is simple: plots of observed data are fitted by eye, either literally freehand (e.g. Stroh, 1971), or with the aid of flexible or 'French' (fixed) curves (e.g. Cannell and Cahalan, 1979). From the resulting lines fitted values of the observed data can be read off for insertion either into the classical formulae (Chapter 2) or for use in the calculation of instantaneous fitted values of the derived quantities.

4.3.2 In the classical approach

An example of the use of freehand curves as a precursor to the calculation of classical growth analytical quantities was provided by Williams (1946). In order to reduce the need for frequent estimates of leaf protein content in an experiment on the nutrition of *Phalaris tuberosa* (canary grass) a freehand calibration of leaf protein against total dry weight was prepared (Fig. 4.3). From this calibration, done separately for each of four nutrient treatments, values of leaf protein content corresponding to the various measures of total dry weight were read off and inserted into the classical formula for the calculation of mean unit leaf rate on a leaf protein basis (section 2.2.4).

In a more limited way, Goodall (1950) used the method to fill in a troublesome corner in his experimental technique. The weights of mature leaves of *Theobroma cacao* (cacao) seedlings were determined from their linear dimensions and the reciprocal of specific leaf area (L_W/L_A – found to be relatively constant) but in immature leaves no clear relationship of this type could be

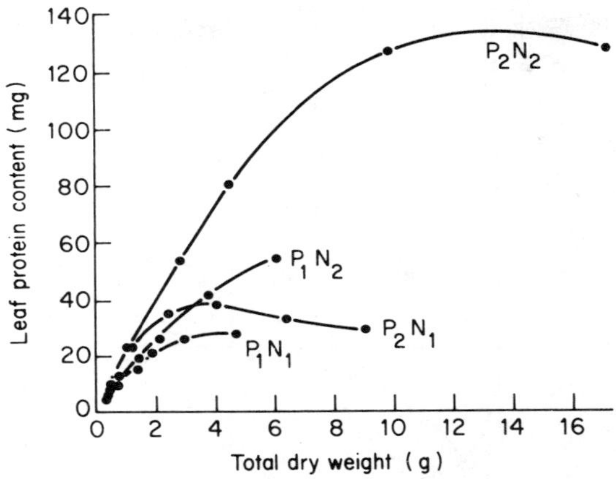

Fig. 4.3 Calibration curves for leaf protein content versus total dry weight in *Phalaris tuberosa*. The plotted points are the observed data and the freehand curves are used for interpolation. Four treatments are shown: these involve low and high levels respectively of phosphorous, P_1 and P_2, and also of nitrogen, N_1 and N_2 (data and curves from Williams, 1946).

found. Freehand curves of dry weight against linear dimensions were therefore employed and interpolations from these again served the classical formulae.

4.3.3 In the derivation of instantaneous values

Tinker (1969) in a study of rates of potassium and water uptake by roots of *Allium porrum* (leek) fitted freehand curves to plots against time of total potassium content, K, total root length, R_L, and cumulative transpiration per plant, CT (a single, two-character symbol), in order to determine, firstly, the 'nutrient ion flux' into the root, $(1/R_L)(\mathrm{d}K/\mathrm{d}T)$, more familiar to us as specific absorption rate (section 2.2.7) and, secondly, water flux into the root $(1/R_L)(\mathrm{d}CT/\mathrm{d}T)$ (both converted onto a root volume basis by multiplying by the appropriate factor). To do this, freehand tangents to the curves for K and CT were taken, giving $\mathrm{d}K/\mathrm{d}T$ and $\mathrm{d}CT/\mathrm{d}T$. Empirical rather than freehand functions were decried because 'With only 4 harvests curve fitting cannot be very accurate.' Later reading of this chapter will show that with few harvests curve fitting can be all too accurate but for the moment readers must balance for themselves this supposed gain in accuracy against the problem of drawing freehand tangents.

Two publications from Czechoslovakia have probably taken the technique of derivation of instantaneous values from freehand curves as far as it will go. In the first, Dykyjová, Ondok and Přibáň (1970) present a convincing figure showing freehand curves fitted to (untransformed) W and L_A in natural stands of *Phragmites communis* (reed) at Trebon (Fig. 4.4a). The $\mathrm{d}W/\mathrm{d}T$ was

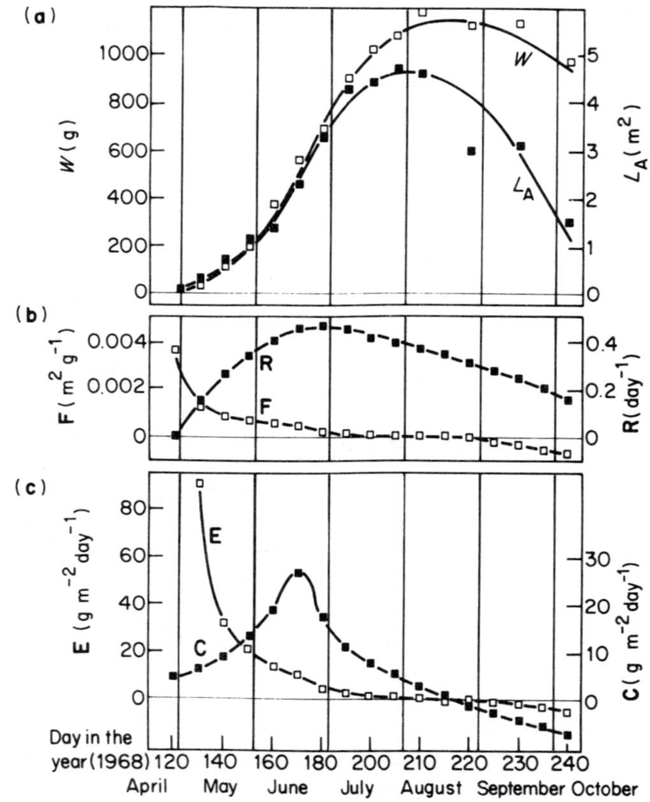

Fig. 4.4 (a) Freehand curves fitted to above-ground biomass (W) and leaf area (L_A), each assessed as totals for 0.5 m^2 quadrats in a natural stand of *Phragmites communis* at Trebon (Czechoslovakia) in 1968; **(b)** relative growth rates (**R**) and leaf area ratios (**F**) derived graphically from **(a)** in the manner described in section 3.4.4; **(c)** crop growth rates (**C**) and unit leaf rates (**E**) derived in the same way (re-drawn from data and analyses given by Dykyjová, Ondok and Přibáň, 1970).

determined graphically at various times to obtain crop growth rate (equation 3.26) and instantaneous, interpolated values of W and L_A were divided into this quantity to obtain relative growth rate (equation 3.9) and unit leaf rate (equation 3.27). In addition, leaf area ratio was derived by dividing instantaneous, fitted values of L_A by those of W (equation 3.10). Plots of these four derived quantities are given in Figs 4.4b and 4.4c. Much of the theory behind this technique is dealt with by Ondok and Květ (1971) who demonstrated how to take tangents to freehand curves and also discussed the value of these curves in the calculation of integral values of **R**, **E**, **C** and **F** as alternatives to the harvest-interval means in Table 2.6. Using a similar methodology to Dykyjová, Ondok and Přibáň (1970), Szeicz, van Bavel and Takami (1973) obtained instantaneous values of crop growth rate and unit leaf rate from freehand curves of W/P and L_A/P versus time. Their material was *Sorghum bicolor* (sorghum, cv. RS 160) grown in Texas. Hannam (1968) provided an example of the same kind of techniques applied to the growth analysis of individuals. Freehand curves were fitted to the logarithms of the lengths of successive leaves in *Nicotiana tabacum* (tobacco). From these, instantaneous values of relative growth rate were derived. Similar approaches were adopted by Hogetsu *et al.* (1960), Elmore *et al.* (1967), Ješko (1972), Williams (1975), Attiwill (1979) and Chatterton and Silvius (1979).

4.4 Fitted curves in the classical approach

This procedure should be self-evident from its title: mathematical functions are fitted to primary data and from these functions fitted values of the primary data are abstracted for use in the classical harvest-interval formulae (Chapter 2). It may seem perverse to come so close to a fully mathematical approach and then revert to classical methods but there may occasionally be slight advantages. For example, a single mean value of **R** or **E** may be needed for a period of growth because the experimenter has no room to present anything more elaborate or because he has changed his mind about the presentation of his results since the experiment was designed. But in the absence of such trivial requirements the procedure has little to recommend it.

This approach sometimes appears in strange mixtures of analytical techniques, for example, Brewster and Tinker (1970) calculated inflow of water to the roots of *Allium porrum* (leek) by means of regressions fitted to transpiration loss versus time, but when faced with the problem of deriving specific absorption rate, they rejected this fully-functional approach and also admitted that

> 'Drawing tangents to the curves to get uptake rates . . . was rather inaccurate'

(section 4.3.3), settling instead for the classical formula (equation 2.7) with values of plant nutrient content derived from exponential curves. Květ (1978) and Mauney, Fry and Guinn (1978) proceeded similarly.

Hammerton and Stone (1966) had good reason to adopt this approach since they were interested in a specific comparison between it, the classical approach and a fully-functional method. Their paper holds much of interest for the student of plant growth analysis and will be referred to again (section 8.2.3).

4.5 The practical background to the functional approach

The reaction to Blackman's (1919) paper among other physiologists working on problems of plant growth was rapid: the succeeding two decades saw many attempts at deriving Blackman's 'efficiency index' from simple mathematical functions fitted to series of experimental observations. Much of this work stemmed from the then Department of Plant Physiology and Pathology at Imperial College, London. A selection of publications from this period came from Gregory (1921), Vyvyan (1924), Gregory (1928a, b), Ashby (1929, 1930), Heath (1932), Hicks and Ashby (1934), Ashby (1937), Portsmouth (1937) and Heath (1937a, b). Among the growth functions employed were the monomolecular function (equation 3.3, and section 6.2), the autocatalytic or logistic (section 6.3):

$$W = a/(1 + be^{-cT}) \tag{4.1}$$

where a, b and c are constants and e is the base of natural logarithms, and both the simple exponential (linear fits to the logarithms of the dependent variate) and more complicated polynomial exponentials (Causton, 1970; non-linear polynomial fits to the logarithms of the dependent variate).

Although much of this research was at least partly motivated by mechanistic ideas which are now superseded, the papers still make interesting reading because it was then the agreeable fashion to present bodies of primary data in some detail. Many of the above-listed functions, too, are still in use as empirical models. For example, some of Gregory's (1921) figures showing autocatalytic fits to leaf area in *Cucumis sativus* (cucumber, cv. Butcher's disease resister) can scarcely be bettered by any more modern technique (Fig. 4.5). We see once again that this division of interests between

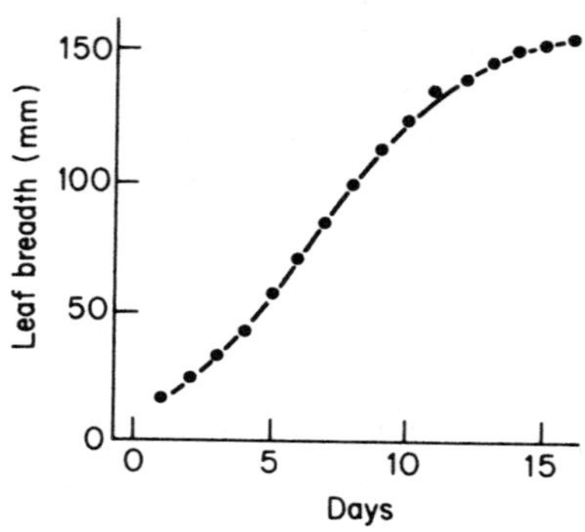

Fig. 4.5 The growth in breadth of a single leaf of cucumber. The fitted function is the logistic (equation 4.1, parameter values unspecified; data and analysis from Gregory, 1921).

the mechanistic and empirical modeller of plant growth (section 3.2) is far from absolute. Both in this early work and, as will be seen, in more modern empirical studies the mechanistic viewpoint cannot be entirely disregarded.

4.6 The choice of function

In strictly objective terms, if 'accuracy in the fit achieved becomes the primary aim' (Richards, 1969) then the identity of the empirical model itself is of no importance whatever so long as it does the best possible job in representing the primary data. While it may not be possible to argue with this viewpoint on theoretical grounds, the experimenter who adopts it as his only *modus operandi* will encounter practical difficulties from two sources.

Firstly, it is an inescapable fact that the more complicated the mathematical function (that is, the more parameters it contains) the more closely it will fit any given series of experimental observations (that is, the smaller will be the deviations between the fitted function and the original data). To take some simple examples, a first-order polynomial (two parameters) will fit two sequential data points exactly, a second-order polynomial (three parameters) will fit three sequential points exactly, and so on. If the data points are not individuals or means of several experimental observations but arrays of replicates, each normally distributed in the Y-dimension, then these exact fits will be to the means of these arrays. By this process even extensive series of observations can be fitted perfectly so long as the empirical model is complicated enough in structure. To take an example of how this can happen, I return to the data on the growth of *Zea mays* (maize) from Kreusler, Prehn and Hornberger (1879) (first quoted in Table 1.1). In theory, the first eleven points in this set of data can be fitted perfectly by a tenth-order polynomial (which has eleven parameters, including the constant term):

$$\log_{10} W = a + b_1 T + b_2 T^2 + \ldots + b_{10} T^{10} . \tag{4.2}$$

When attempting this, the curve-fitting program (BMD05R from Dixon, 1973) reached no higher than a polynomial of the ninth order because the computational accuracy was not sufficient to make further progress – itself an indication of the absurdity of the exercise. The analysis of variance for this ninth-order polynomial exponential is given in Table 4.1 and the curve itself is shown fitted to the data in Fig. 4.6.

One can see from Table 4.1 that the higher-order terms, say from the sextic onwards, remove a negligible proportion of the total variance. From Fig. 4.6 it is evident that the fit achieved is virtually as close as that which might have been obtained by joining the data points by a series of straight lines (Fig. 2.1).

There are objections to such an empirical model on three grounds: (1) the computation is laborious, if not impossible, without full-scale computing support; (2) little or no smoothing of the data is done – every small inflection is reproduced (this may or may not be a disadvantage, but it is uncommon for the empirical modeller to require no smoothing *at all*) and (3)

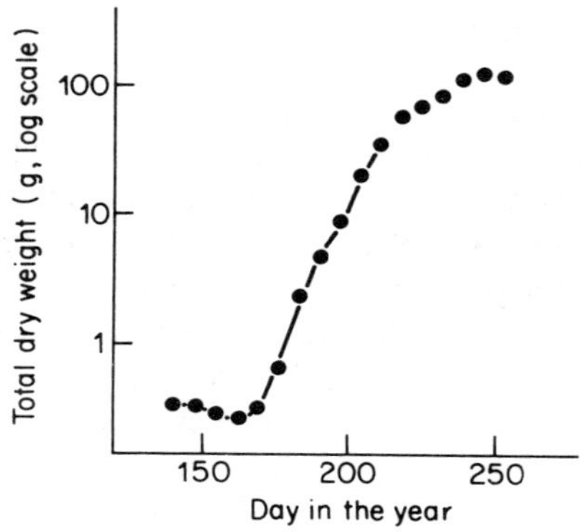

Fig. 4.6 Data from Kreusler *et al.* (1879) on the growth of 'Badischer früh' maize in 1878 (Table 1.1); a ninth-order polynomial exponential fitted to the first eleven mean dry weights (see Table 4.1).

Table 4.1 An analysis of variance for a ninth-order polynomial fit to the logarithms of the first eleven points in the data set of Kreusler, Prehn and Hornberger (1879). See also Table 1.1 and Figs 2.1 and 4.6.

Source	Degrees of freedom	Sum of squares
Linear term	1	5.71586
Quadratic term	1	0.65630
Cubic term	1	0.12765
Quartic term	1	0.03692
Quintic term	1	0.03752
Sextic term	1	0.00025
Seventh-order term	1	0.01549
Eighth-order term	1	0.00168
Ninth-order term	1	0.00003
Deviation about regression	1	0.00458
Total	10	6.59628

in complicated non-polynomial models the estimation of derived quantities such as first-order differentials is more difficult than in the case of simpler models and, most importantly, the statistical estimation of errors in complicated models, and in their derivatives, becomes extremely difficult (and even in high-order polynomials such errors tend to be very large). For these reasons the experimenter should prefer the simplest model possible that is consistent both with the degree of smoothing required or desired and his own mechanistic insights into the process that is being modelled. Even on strictly objective grounds, such high-order polynomials cannot be justified by every test since, although the ninth-order fit has the lowest possible residual sum of squares, its F-ratio (regression mean square/residual mean square), which is 190.0, is by no means the highest in the polynomial series. For this set of data that distinction belongs to the seventh-order polynomial, with an F-ratio of 449.0. (See also p. 112.)

A second reason why empirical models which approach perfection statistically may yet be unsuitable representations of the data, is that they may not satisfy the aforementioned mechanistic insights of the experimenter. To take a simple example, if early growth, such as that of frond number in *Lemna* sp. (duckweed) under non-limiting conditions, is known to be exponential then the exponential model (first-order polynomial exponential) is the only model which the experimenter should fit. He may quite justifiably take the view that any deviations from exponentiality in his data are of no real importance and deserve to be smoothed out – despite the fact that if such deviations were to be respected some other model would be a better fit on purely statistical grounds. Such a course was taken by Gregory (1921). Figure 4.7, from his paper, shows a clear sinuosity in the growth of his population of cucumber leaves. This is related to the unfolding and growth of the individual leaves themselves, making L_A only a quasi-continuous variate. Nonetheless, analysis by means of the exponential function was adhered to and these individual subfeatures of the curve were submerged into an overall trend.

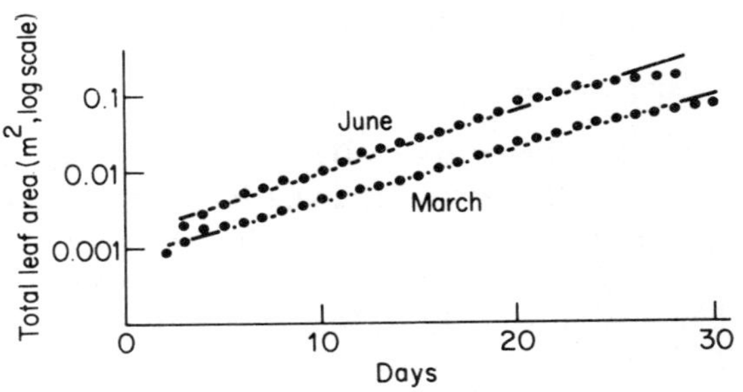

Fig. 4.7 Progressions on time for total leaf area in cucumber. Two series of glass house data are shown, one for plants started in March 1917 and one for plants started in June 1917. The fitted regressions are first-order polynomial exponentials: $\log_e L_A$ (March) $= 0.92 + 0.069T$; $\log_e L_A$ (June) $= 1.15 + 0.085T$ (data and analyses from Gregory, 1921).

As plants grow the proportion of purely structural material that they contain increases. This is an inescapable physical constraint because for each linear increment in height the surface area of the plant increases by something like its square and the weight of the plant by something like its cube. This simplistic view of growth is not undermined by its being inexact. For these reasons relative growth rate cannot remain constant for long and eventually must show a decline as more and more of the plant's material becomes purely structural and therefore incapable of providing further increases in dry weight (section 2.2.4). For these perfectly valid reasons some experimenters prefer models which not only allow the possibility of a decline in RGR with time but, by their own nature, expect it. Such a model is the second-order polynomial exponential advocated by Hurd (1977). Hurd also argued that

> 'statistical methods [that is, empirical functions chosen on purely objective grounds] cannot be used unmodified for experiments involving comparisons between artificially-lit cabinets'

because they do not necessarily provide

> 'the similar patterns of growth that would be expected between closely related steady-state treatments.'

The hard-nosed might argue that if experimental data do not conform to one's theoretical expectations then one either needs to obtain data that do, or to alter the expectations themselves. But Hunt (1978a), although disagreeing ('Hurd's suggestion that a quadratic regression might be used throughout . . . needs to be treated with caution') took a less extreme viewpoint. While suggesting that high-order polynomials can be avoided by increasing the stringency of the tests required for their acceptance, the author warned that, in the body of Hurd's data under discussion,

> 'at no time is a single model unquestionably the best.'

In practice, this would mean rejecting Hurd's second-order models in cases where no fit other than the first-order could be justified statistically. This move towards the greatest possible simplicity in the choice of growth functions was supported by the author's own experience with the higher-order polynomials (Hunt and Burnett, 1973; Hunt and Parsons, 1974) and by the work of Nicholls and Calder (1973).

Kreusler's data on the growth of maize (Fig. 1.1b) exhibit a trend which is clearly more complex than that of either the first- or second-order polynomial exponential. Any analysis of such sets of data by way of these models alone would fail miserably to represent the real situation (see sections 5.2 and 5.3). And yet, this pattern of growth is, as has been stated (section 1.2), exceedingly common. With this type of example in mind, Hughes and Freeman (1967) suggested that a third-order polynomial exponential model might be of general utility. In one of its forms this model can reproduce symmetrical, S-shaped curves in the logarithmically transformed data with some accuracy and the degree of sinuosity (but not its symmetry) can be varied infinitely. Superficially, it is an attractive model for data such as Kreusler's (but see section 5.4); the difficulty arises when it is applied to data in which the underlying trends are simpler than those of Kreusler's set. This disadvantage was not seen by Hughes and Freeman, for in 1971 A. P. Hughes, in a personal communication to the author, wrote

> 'The reasons why I like our cubic programme just the way it is are:– . . . the cubic fit allows two points of inflexion [*sic*] so that we can have a sigmoid with a final decline. Many other S-shaped curves are restricted to be asymptotic [and] in practice we have found over many experiments that it works. If the cubic term is not significant it will be small and vice versa. It therefore does not help much to go back to the quadratic if the cubic term is not significant. . . . I shall be interested to see whether my general feeling that the cubic can be left in and have little effect is true. I am certain that the standard error of the fitted values and derived data will not get much smaller, and that the corresponding fitted values of the quadratic and cubic fits will not be significantly different.'

The differences, particularly in the derived quantities, turned out to be substantial (Hunt and Parsons, 1974).

To take stock now of the problems involved in the choice of empirical model, it may be stated that this choice is determined by two requirements which often conflict: statistical exactitude and what Hurd (1977) called the experimenter's 'biological expectation'. If the experimenter can justify either of these as being of overriding importance then the other can be ignored. But in the more usual case where both are given some credence, the experimenter himself can be the only arbiter.

In this discussion of what might be called 'statistical versus biological considerations' in the choice of model, I have deliberately confined myself to the first three orders of polynomial exponential because they illustrate the problems so neatly. But other polynomial exponentials, and other curves not of this type at all, still face the same difficulties: Is a certain number of parameters needed or not? Do they achieve good statistical fits? Do they satisfy any mechanistic ideas that the experimenter may have? I shall return to these problems in Chapters 5, 6 and 7 when dealing with further examples of fitted growth curves.

4.7 Statistical considerations

4.7.1 The practicalities of curve fitting

In the remaining pages of this chapter I shall deal with more statistical aspects of curve-fitting but, because this is not primarily a statistical text and, more importantly, because I am not a statistician, I gladly pass much of this task into more capable hands. I shall deal only with two specific problems, each having a peculiar relevance to plant growth analysis: that of transformations of primary data and that of variability in these primary data. But firstly, some guidance is called for in the practicalities of curve-fitting.

If the experimenter has decided that a first-order polynomial exponential is an adequate model then several suitable guides to its derivation are available, for example, Bailey (1964, p. 91), Mather (1964, p. 109), Clarke (1980, p. 92) and Parker (1978, section 8.6). Hunt (1978a, p. 41) has given a detailed example of the calculation of mean RGR and its standard error from the linear regression of the logarithms of dry weight using the data on the growth of *Holcus lanatus* (Yorkshire fog) which are presented in the next section (Fig. 4.8). In an older publication concerning the growth of the main stem of *Gossypium* spp. (cotton), well worth obtaining despite its relatively obscure source, Afzal and Iyer (1934) gave a most detailed account of the calculation of mean RGR and its errors from the simple exponential equation. Extensive figures and tables of primary data were presented and the intermediate quantities in the calculation (the various sums of squares) were also displayed at length. Finally, full statistical details were given for comparing values of mean RGR by Fisher's method of pooled variances (described most recently by him in Fisher, 1970, p. 140).

If the use of a higher-order polynomial is a possibility, or a polynomial exponential (that is, using logarithmically-transformed data), then Mather (1964, p. 129), although employing orthogonal polynomials (section 5.1), may be useful. Using Ashby's (1937) data on the growth of *Lycopersicon esculentum* (tomato), Mather fits first-, second- and third-order polynomials to $\log_e W$ versus time. The calculations are laid out in detail and Mather includes an extensive discussion on criteria for the choice of polynomial. More advanced texts dealing with curve-fitting are those of Draper and Smith (1966) and, dealing with the special problems posed by phenomena like seasonal trends, of Kendall (1973). Other aspects of the fitting of polynomial exponentials have been covered by Hammerton and Stone (1966) and an analysis of the exponential growth equation as a predictor of growth (from a knowledge of photosynthetic and respiratory data) has been provided by McKinion, Hesketh and Baker (1974).

Outside the field of polynomial regression, basic facts about the Gompertz curve have been collated by Winsor (1932) and about the logistic by Fresco (1973). Richards (1959) has devised a function traditionally written as

$$W = a(1 + b\mathrm{e}^{-cT})^{-1/d} \qquad (4.3)$$

where a, b, c and d are constants and e is the base of natural logarithms (see section 6.5). This function has been considered by many to have great potential as an empirical model for plant growth and has been discussed by Nelder (1961, 1962) and Causton (1969, 1970).

Among general publications dealing with the use of regression techniques in plant growth studies are discussions by Petersen (1977) on the role of regressions *vis-à-vis* multiple comparison procedures (such as Duncan's multiple range test (Duncan, 1955)) and by Mead (1971) and Mead and Pike (1975) who point out with some astringency common misconceptions or areas of ignorance which many biologists possess in the field of regression analysis. Zar (1967) has written some notes for experimenters who want to change the units in which their primary data are expressed *after* performing regressions.

For those wishing to 'survey the field' before choosing which regressions to try on their data, there are useful comparisons of a range of growth functions contained in Ondok (1971b), Causton (1977), Landsberg (1977) and Milthorpe and Moorby (1980). Comment of a more statistical nature has been provided by Gladilin and Orlovskii (1973), Kowalski and Guire (1974) and Sandland and McGilchrist (1979). Further references are provided in later sections, as required.

4.7.2 The transformation of primary data

Three important prerequisites exist for regression analysis. Firstly, that an independent variable, X, should be measured without, or virtually without, error; secondly, that the distributions of replicated Y values at each X should be normal; thirdly, that the variance of these subpopulations of Y should be uniform throughout the analysis and not change in magnitude with X. Unless these three conditions are observed no simple method of calculating the regression can be successful. The first condition is not difficult to satisfy: in plant growth analysis time can be measured with much greater accuracy than the other quantities. It is the second and third conditions which nearly always need special attention. The usual way around this problem is to transform the values of the primary data and then perform the regressions on these transformed data. For this purpose the logarithmic transformation is almost standard. There is no reason why logarithms to any particular base need be used, but since logarithms to the base e (natural logarithms) provide a convenient starting point for the calculation of many of the derived quantities (sections 3.4.3, 3.4.5 and 3.4.7) they are the conventional choice. The importance of this transformation may be appreciated by comparing Fig. 4.8a with Fig. 4.8b, for once this effect can be visualized, no further words are needed. Experimenters in doubt as to whether or not a transformation of data is required should apply Bartlett's test for homogeneity of variances (e.g. Bailey, 1964, p. 114).

In the literature, publications occasionally appear which obviously reveal a need for such transformations (see the 'Comments' in the Tables which follow). Included here would be those of Allison and Watson (1966), Goodman (1968) and Allison (1969, 1971). Figure 4.9, taken from Allison (1969), shows the problem clearly. In a study on the growth of *Zea mays* (maize) at Salisbury, Rhodesia, Allison was compelled to ignore all data obtained before the ten-week harvest, and also leaf area data obtained after the eighteen-week harvest, when fitting quadrátic polynomials to the untransformed data. Classical methods of analysis were used for these periods.

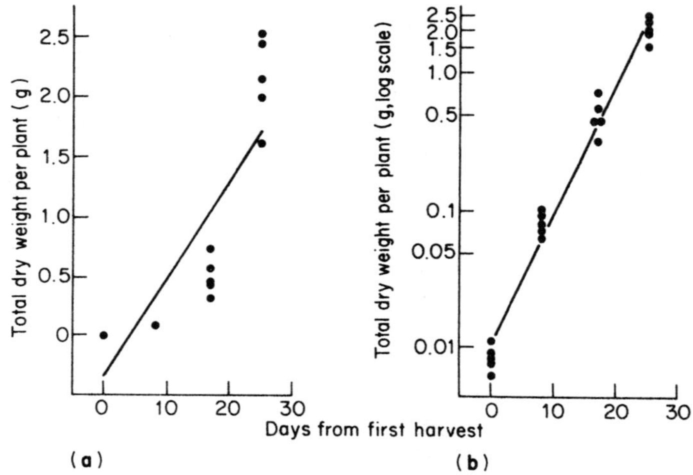

Fig. 4.8 (**a**) A first-order polynomial fitted to a plot of total dry weight versus time in young seedlings of *Holcus lanatus*. The first two harvests each contain five replicate plants but these cannot be shown individually when plotted on this scale. The standard error of slope in this regression is 13.7 per cent of the value of the slope itself; (**b**) the same data and the same analysis after transformation to natural logarithms (creating a first-order polynomial exponential fit). The standard error of slope is now only 2.9 per cent (data of Grime and Hunt, 1975; also dealt with by Hunt and Parsons, 1974 and Hunt, 1978a).

4.7.3 The variability of primary data

Elias and Causton (1976) discussed the question of which degree of polynomial should be fitted to logarithmically transformed raw data, when employing the functional approach to plant growth analysis. By artificially manipulating data on the growth of several species these authors showed that when growth curves were fitted to data of low variability, unrealistically high degrees of polynomial were quite likely to be selected if purely statistical criteria alone were used. When variability was high, however, the 'best fit' was more likely to be a lower-order polynomial. Elias and Causton concluded that regressions should be fitted using harvest mean values where

> 'the test of adequacy of fit is independent of the underlying population variability',

even though this leads to a loss of information. For example, in one of Elias and Causton's comparisons, *Blackstonia perfoliata* (yellow-wort) grown in a glasshouse was harvested on thirteen occasions over its first forty-six days of growth. A fully replicated plot of the total dry weights of the plant was best fitted by a fifth-order polynomial exponential, whereas the same trend artificially enhanced in variability by removing the four observations nearest

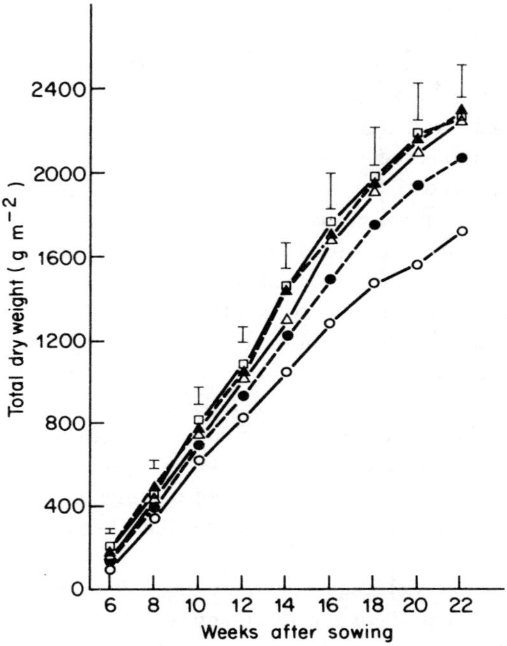

Fig. 4.9 Growth curves for total dry weight in 1 m^2 plots of maize grown at Salisbury, Rhodesia. The various symbols represent five different populations, grown concurrently. Natural variability in these data, as revealed by least significant differences (vertical bars, $P<0.05$), increases markedly as growth proceeds (data from Allison, 1969).

each harvest mean justified only a quadratic fit. Clearly this is a case where the experimenter's 'biological expectation' should override statistical exactitude (section 4.6).

4.8 In conclusion

The philosophical and statistical matters that have been covered in the last two chapters can vary from being of overriding concern to some experimenters to being of no concern at all to others. As so often, the middle road is the best way with a balanced appreciation of the problems that confront workers in this field being essential, but without obsession to details of rationale, experimental procedure or statistical technique. The experimenter who can see all of these problems equally is best placed to advance his subject if plant growth analysis is involved.

4.9 A forward look

The next chapters will move on to what is in effect a classified list of case-histories in the functional approach. By displaying the considerable range of existing published work it is hoped that new experimenters may be able to pick and choose among the models in the light of others' experience. Separate chapters will be devoted to the use of polynomial functions, asymptotic functions and of the more unusual or complex functions.

For each function within these broad families, its mathematical structure will be described in simple terms and sources of further information will be listed. With particular reference to the utility of the function in plant growth analysis we will examine common or useful variants of the function and see graphically the consequences for relative growth rate, and other derivates of adopting each of its particular forms. Existing uses of the function in the literature will be surveyed and exemplified and, from time to time, we will see how Kreusler's data fare (Table 1.1) when approached in each particular way. We may also see instances of errors of commission and omission: cases where the function was used and should not have been, or was not used but might have been.

Finally, the whole field will be drawn together in the last chapter when we will compare the classical and the functional approaches to plant growth analysis, examine the biological relevance of fitted growth functions, compare the various types of function and ask what, if any, is the future for this type of activity within plant science in general.

5

Polynomial functions

5.1 Introduction

In a polynomial function the dependent variate is expressed as a function of more than two terms. The title is particularly applied to expressions involving sums of independent terms which are each different powers of the same primary variable. The name is a hybrid, derived from the Greek πολύ (= many) and the Latin *nomen* (= name or term). It was originally coined to complement the adjective *binomial* (= consisting of two terms).

The title is generic and the genre includes, in theory, an infinite number of variants. Although, strictly speaking, it must be outside the definition of a polynomial because it contains only two terms, the so-called first-order (or first-degree) polynomial, which in the context of plant growth analysis may be exemplified

$$W = a + bT \tag{5.1}$$

is generally held to be the starting point of a series which ascends in complexity through the addition of higher powers of T.

In its logarithmically-transformed form, the first-order polynomial (or polynomial exponential in the terminology of Causton, 1970) is

$$\log_e W = a + bT \tag{5.2}$$

because $W = \exp(a + bT)$. The series then ascends through

$$\log_e W = a + b_1T + b_2T^2 \tag{5.3}$$

and

$$\log_e W = a + b_1T + b_2T^2 + b_3T^3 \tag{5.4}$$

to

$$\log_e W = a + b_1T + b_2T^2 + b_3T^3 + \ldots + b_nT^n. \tag{5.5}$$

Members of this polynomial series take their names from the highest power of the independent variable (see Table 4.1). As we have seen (section 4.7.2), the logarithmically-transformed series is of more interest and value to plant growth analysis than the untransformed series. It is certainly true to say, too, that for us the most important member of the series is the first, and that relevance and usefulness decline sharply as the series ascends, reaching zero well before the order of polynomial rises into double figures. For this reason we will deal with polynomial functions in this chapter under four headings alone: linear, quadratic, cubic and higher order, concluding with a survey of methods which deal with different orders simultaneously in multi-order 'packages'.

All polynomial functions share a number of convenient mathematical and statistical properties, which have combined to make this group of models the most popular of all for experimenters faced with the problem of fitting plant growth curves. Unlike the asymptotic functions dealt with in the next chapter, parameter values, derivates and the variances of both are, for polynomial functions, straightforwardly accessible through an established corpus of theory. Briefly, this means that exact and rigorous methods exist of calculating estimates of the various parameters independently of one another. Draper and Smith (1966, p. 44ff) included a most suitable general guide to the theory and practice of polynomial regression and Causton and Venus (1981, pp. 65–85) provided a whole chapter on linear regression theory, in preliminary support of their later exposition of the Richards function, a non-linear expression. Their chapter covers such matters as estimation of parameters and their variances and includes notes on the two most practical alternative ways of obtaining these: least squares estimation and maximum likelihood estimation.

After weighing all of the evidence presented in the remainder of this chapter, the experimenter committed to polynomial functions will have to decide on practicality of three different courses of action: (1) to obtain from an existing worker in the field a complete computer program written in a high-level language and capable of providing all of the required parameters and derivates; (2) to employ an existing commercial regression 'package' (such as that of Dixon (1973) or Nie *et al.* (1975) as updated by Hull and Nie (1981)), which may need supplementation since it will not go as far as is required, for example, in providing variances and derivates; or (3) to construct his own scheme of analysis, working from first principles and guided by the sources cited in the preceding paragraph.

Finally, if full-scale computing facilities are either unavailable or 'unpalatable' there are the special, although limited, regression possibilities available through the method of orthogonal polynomials. Briefly, if the dependent variate is unreplicated, or present as one mean value per harvest, and if the harvesting has been equally spaced, then it is possible to determine from the experimental data and from specially tabulated values the regression coefficients of a range of the simpler polynomial functions. For example, Fisher and Yates (1963, Table XXIII) supply values for fitting polynomials of up to the fifth order to up to 75 data points. They provide references to other works which variously supply tables for polynomials of up to the ninth order and for up to 104 data points. Erickson (1976) has dealt with the use of orthogonal polynomials in fitting plant growth curves and, in the context of the method of running averages (see the present section 7.3), has provided tables for first- or second-order smoothing polynomial fits to three or five points, for second-order fits to seven points, and also for first derivatives (absolute or relative growth rates, section 3.4) in the form of first- or second-order fits to three points, first-, second or third-order fits to five points, and second-order fits to seven points.

5.2 First-order polynomial

5.2.1 The function and its properties

As we have seen, the first-order polynomial or 'linear regression' is, for total dry weight,

$$W = a + bT \tag{5.1}$$

or, as a first-order polynomial exponential

$$\log_e W = a + bT \tag{5.2}$$

with parameter a being expressed in logarithmic units in equation 5.2. From equation 5.1 an absolute growth rate may be derived

$$\frac{dW}{dT} = b \tag{5.6}$$

and, from equation 5.2, the relative growth rate is

$$\frac{1}{W} \cdot \frac{dW}{dT} = \frac{d(\log_e W)}{dT} = b \ . \tag{5.7}$$

Equation 5.2 is frequently referred to as the 'exponential equation', although that title more properly refers to the integrated form of equation 5.7:

$$W = ae^{bT} \ . \tag{5.8}$$

Parameter a is the value of W or $\log_e W$ when $T = 0$ and is thus indicative of the size of the growing system at the origin of the time scale chosen for the analysis, wherever that may lie. Parameter b is the rate of increase of W or $\log_e W$ with T, the absolute or relative growth rate, **G** or **R**. The latter may also be called the rate of exponential increase, rate of compound interest, specific growth rate (for dry weight only) or efficiency index. Figure 5.1 shows in a stylized way three variants of equation 5.2: in part (a) we see a positive value of b which leads to a steadily increasing value of $\log_e W$ and a stationary, positive value of **R**; in part (b) parameter b is zero, indicating a finite but stationary value of $\log_e W$ (in fact, a) and a zero **R** throughout; in part (c) parameter b is negative, $\log_e W$ declines steadily and **R** is stationary and negative. All of this could, of course, be repeated in terms of W and **G** were untransformed primary data to be used.

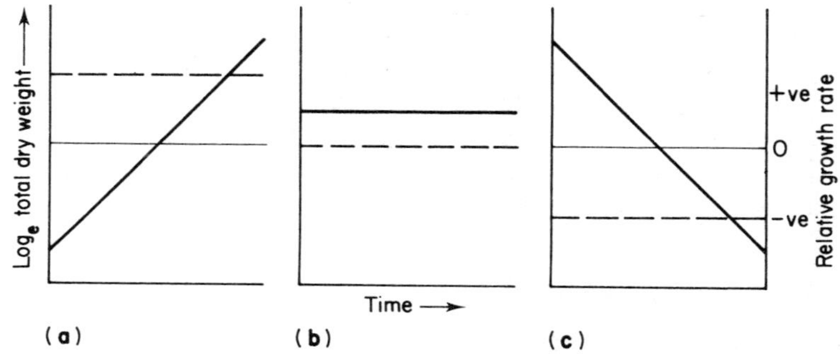

Fig. 5.1 Specimen first-order polynomial exponentials (equation 5.2) showing the progressions of $\log_e$ total dry weight (——) and of relative growth rate (– – –); **(a)**, b positive; **(b)**, b zero; **(c)**, b negative.

The first-order polynomial exponential (equation 5.2) is alone among the growth curves in this volume in possessing, on suitable occasions, a sound mechanistic basis. Where plant growth, in single organs or in very young whole individuals, proceeds by a series of equal divisions of cells at regular intervals, then the number and total weight of cells increases geometrically and exponentially with time. The dry weight of this wholly meristematic system thus increases in a log-linear fashion and the first-order polynomial exponential is an entirely appropriate model. However, wholly meristematic systems cannot sustain themselves indefinitely and the consequent division of labour among cells and tissues means that the applicability of the first-order polynomial exponential will decline as time goes by. Hence, its greatest use in modelling the growth of higher plants has been in the earliest phases where the assumption of exponentiality cannot be discounted *within the limits of the information supplied by the experimental data*. In the 384 regressions performed by Grime and Hunt (1975) while screening for **R** over the period 14–35 days after germination, exponential growth at $P<0.05$ was seen in 210 cases.

In addition to cases in which the first-order polynomial exponential is the most desirable of all growth functions, there are at least three instances in which it may be 'forced' onto experimental data in order to achieve some particular purpose. To illustrate these, we turn to three specific examples involving the positive, zero and negative progressions of **R** depicted in Fig. 5.1.

Where the plot of, say, $\log_e W$ on T is clearly non-linear it may still be valid to fit the linear model if the experimenter needs a single value of **R** for comparative purposes. Done in this way, **R** is in one sense a mean value, $\bar{\mathbf{R}}$: the average slope of the whole plot of $\log_e W$ on T. In this respect it is quite distinct from the mean slope between initial and final values of $\log_e W$ (equation 2.7) since it takes into account all of the available data at once and not merely

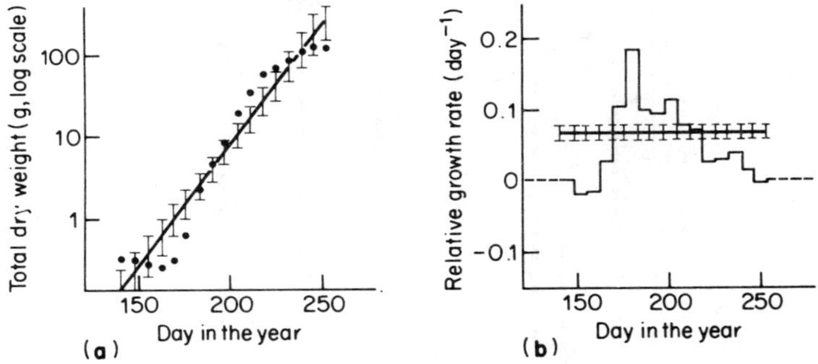

Fig. 5.2 Kreusler's data for maize (Table 1.1) fitted by a first-order polynomial exponential; (**a**) observed values of total dry weight per plant, with the fitted line and 95% limits; (**b**) fitted values of relative growth rate, with 95% limits, shown against the harvest-interval means.

those from the intial and final harvests. Kreusler's data for maize (Table 1.1) are shown fitted by the first-order polynomial exponential in Fig. 5.2, with the resulting progression of **R**. We see in part (a) that the equation is hopelessly lacking in suitability, with up to seven consecutive data points at a time lying to one side of the fitted line. In part (b) we see that the resulting uniform value of **R**, 0.068 ± 0.010 day^{-1}, lies at first above, then below, then above again, the harvest-interval means of $\bar{\mathbf{R}}$ displayed in the background. It is also slightly different from the value ${}_{140-253}\bar{\mathbf{R}}$ calculated from the initial and final harvests, 0.052 day^{-1}. Nonetheless, it confirms in a rough-and-ready way, if this is what is required, that Kreusler's maize displayed an all-in average gain in dry weight of almost 7 per cent per day, plus or minus 1 per cent.

While, in many instances, a relationship such as that exhibited in Fig. 5.1b may be the depressing result of poor growth, careless sampling or high variability, it is desirable on occasion to use the first-order polynomial exponential *to show that* it is improbable that growth has occurred. For example, Hunt (in prep.) wished to demonstrate that young seedlings raised in solution culture and then killed by increasing the level of copper in the nutrient solution to 1000 g m^{-3} could safely be left *in situ,* neither growing nor decaying, for several hours before completing the harvesting operation (this was so that plants could be killed automatically at very frequent intervals, 12 min, in the absence of the experimenter). Forty individual plants harvested hourly for 6 hours following the killing (Fig. 5.3) showed an entirely stationary progression of $\log_e W$ on T ($P < 0.05$). Because of the expanded scale used in Fig. 5.3 to display clearly as many as possible of the forty replicates per harvest, variability may appear at first sight to be unusually high. But this is not so: $\bar{W}$ was 1.17 mg, with 95% limits of + 0.61 mg and – 0.39 mg (asymmetrical following logarithmic

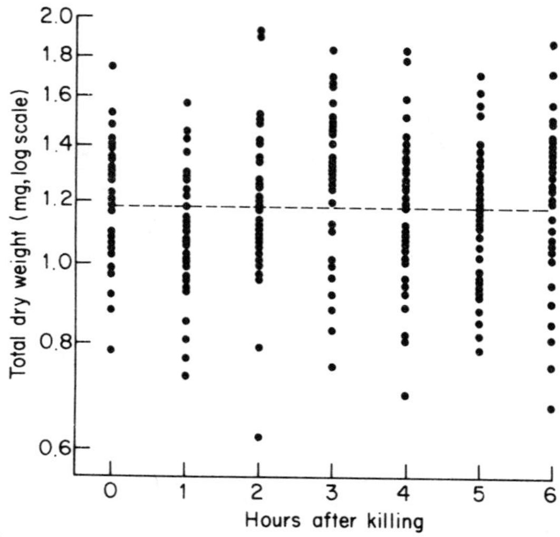

Fig. 5.3 The stationary progression of $\log_e$ total dry weight per plant in *Holcus lanatus* following the establishment of 1000 g m^{-3} Cu^{++} in the rooting medium at time zero. The b coefficient in the fitted regression line is not significantly different from zero at $P < 0.05$.

back-transformation), limits which are quite realistic for a population of seedlings drawn, as this was, from a 'wild' collection.

The final instance to be cited in which a first-order polynomial exponential may be 'forced' onto experimental data, for comparative or other purposes, concerns the situation where $\bar{\mathbf{R}}$ has a negative value and the system under investigation is not growing but decaying (Fig. 5.1c). Solving the equation

$$\text{halving time} = 0.693/\bar{\mathbf{R}} \tag{5.9}$$

in such cases would provide the time interval over which $\log_e W$ decreased by 0.693. Since this number is the natural logarithm of the number 2, the time interval is that over which W has halved, more commonly referred to as a half-life. This concept is familiar enough in microbiology and radiobiology and may, on occasion, be useful in plant growth analysis. For example, the decay of plant material in decomposer cycles often follows a course that may be described by a negative exponential model (Mason, 1977, p. 21). This is essentially equation 2.7 running in reverse (with ${}_1W > {}_2W$) and with $\mathbf{R}$ as the relative *decay* rate. The calculation of a half-life may be a useful summary of events. For example, the negative exponential function, among others, has been used by Gloaguen and Touffet (1980) and Gloaguen, Touffet and Forgeard (1980) in studying the rates of decomposition of the litter of woody species from the forests of Ille-et-Vilaine in Brittany. Hüsken, Steudle and Zimmermann (1978) fitted the same function

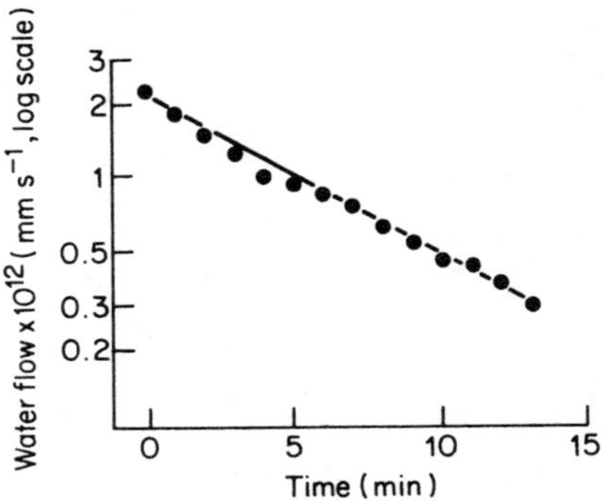

Fig. 5.4 A negative first-order polynomial exponential fitted to sequential measurements of the velocity of water flow across the membrane of a single mesocarp cell in a fruit of *Capsicum annuum* (data and analysis from Hüsken, Steudle and Zimmermann, 1978).

to the velocity of water flow across the membrane of a single mesocarp cell from a fruit of *Capsicum annuum* (green pepper). From the slope of this line (Fig. 5.4) they calculated that the half time for water exchange of the cell following an artificial turgor pressure change of 45 kPa (0.45 bar) was approximately 4 minutes.

So far, in dealing with the first-order polynomial or polynomial exponential we have been concerned solely with the properties of the function for modelling changes in W and $\log_e W$ and in their derivates **G** and **R**. But, of course, it is common in the functional approach to plant growth analysis to proceed still further to quantities of 'type (ii)' and 'type (iii)' (see section 2.1) such as leaf area ratio, **F**, and unit leaf rate, **E**. For these, too, the choice of approximating function fitted to the primary data also has its implications. Though there will scarcely be space to consider these for each and every function listed in the remainder of this text, it is worthwhile presenting an example of a full analysis of this type here, carried through to the calculation of **F** and **E** and their errors, in order to illustrate one very important point: progressions of these later derivates invariably possess greater flexibility than the progressions in the primary variates. This is the outcome of the method of calculation. For example, from section 3.4.3 we saw that while $\mathbf{R}_W$ and $\mathbf{R}_L$ each arise simply from a single approximating function, **F** involves properties of both functions. Even if these are linear, this method of calculation may result in a curvilinear progression in **F**, while **E**, the ratio between a ratio and a rate, may be still more complex in form. Some of this is exemplified in Fig. 5.5 where full fitted data for approximately the first two week's growth of *Holcus lanatus* (Yorkshire fog) are displayed. Though both progressions in the primary data are log-linear ('forced' in the case of L_A, where a quadratic would have been better), and though progressions of

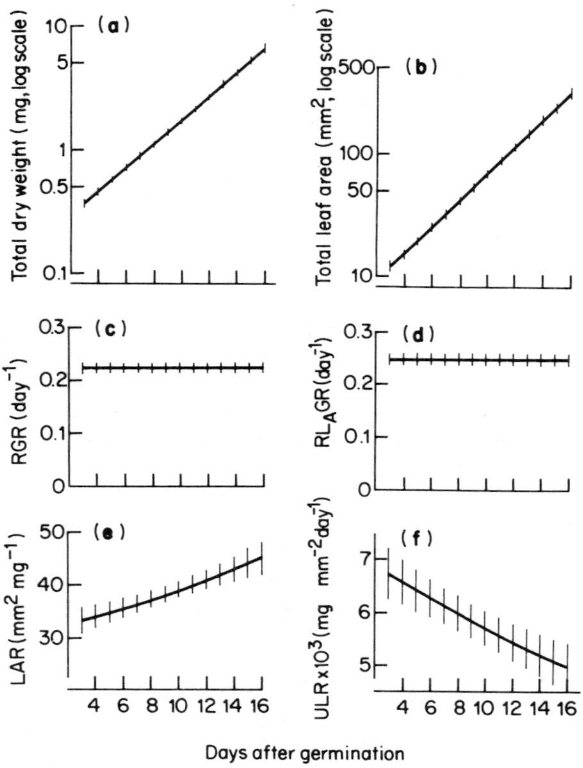

Fig. 5.5 First-order polynomial exponential fits to **(a)** total dry weight and **(b)** total leaf area (both per plant) in *Holcus lanatus;* **(c)** fitted relative growth rates derived from **(a)**; **(d)** fitted relative leaf area growth rates derived from **(b)**; **(e)** fitted leaf area ratios derived from **(a)** and **(b)**; **(f)** fitted unit leaf rates derived from **(c)** and **(e)**; all with 95% limits.

$\mathbf{R_W}$ and $\mathbf{R_L}$ are necessarily simple, **F** and **E** both drift relatively freely along curvilinear courses, in this case the curves of exponential functions.

5.2.2 Case studies

A large selection of work involving fitted first order polynomials is listed in Table 5.1. This selection deliberately excludes examples involving more than two orders of polynomial simultaneously (see section 5.6), or the segmented approach (see section 7.2).

Despite its biological, mathematical and statistical advantages, the logarithmic transformation is by no means compulsory and the calculation of $\bar{\mathbf{G}}$ (or $-\bar{\mathbf{G}}$, the absolute *decay* rate) from untransformed experimental data

Table 5.1 Applications of the first-order polynomial (untransformed primary data) and first-order polynomial exponential (logarithmically transformed primary data).

Author(s)	Species	Time interval(s) (see footnote)	Primary data	Derived data	Comments
Untransformed primary data					
Ashby (1929)	*Lemna minor* (duckweed)	9 days	Frond N		
Ahloowalia (1973)	3 *Lolium* spp.	60 min (e)	Pollen-tube	$\bar{G}$	
Vartha (1973)	*Lolium perenne*, *Poa trivialis* (ryegrass, rough-stalked meadow grass)	9, 17 weeks (e)	ΔW	$\bar{G}$	rates of *re*-growth
Khasawneh (1975)	None, theoretical applications only	–	R_L	$\bar{G}$	Component of uptake model
Webb (1975)	*Pseudotsuga menziesii* (Douglas fir)	4 days (e)	C_f/C_t	$\bar{G}$	Ratio of photoassimilated C to tissue C
de Vries (1976)	*Rosa* sp. (hybrid tea rose)	54 days (g)	Seedling H, flower yield		
Adepetu and Akapa (1977)	*Vigna unguiculata* (cowpea, 5 vars.)	30 days (g)	R_L	$\bar{G}$	Method of Khasawneh (1975)
de Vries and Dubois (1977)	*Rosa* sp. (hybrid tea rose)	32 days (g)	seedling H		
MacColl (1977)	*Saccharum* sp. (sugarcane, 12 vars.)	15–42 weeks (p)	L_A	F, E, L	See Fig. 5.9
Smith, Middleton and and Edmonds (1978)	31 N.Z. crops	21–84 days (e)	Leaf [Na]	$\bar{G}$	$\bar{G}$ contrasted in 'natrophiles' and 'natrophobes'
Wallach, Marani and Kletter (1978)	*Gossypium hirsutum* (cotton, cv. Acala SJ-1)	50–80 days (g)	W m^{-1}, boll W	Inputs to growth model	Time adjusted by degree-days
Raja Harun and Bean (1979)	*Lolium multiflorum* (Italian rygrass, 7 populations)	15–40 days (after anthesis)	$[H_2O]$	$-\bar{G}$	Moisture loss during ripening

Continued

Table 5.1 (*Continued*)

Author(s)	Species	Time interval(s) (see footnote)	Primary data	Derived data	Comments
Lamont and Downes (1979)	*Xanthorrhea preissii*, and *Kingia australis* (grasstrees)	400 years (g)	Total H	$\bar{G}$	Plotted as $T = f(H)$
Silsbury (1979)	*Trifolium subterraneum* (subterranean clover)	33–72 days (p)	W, total N, N_2 fixation CO_2 balance	$\bar{G}$	
Simmons and Crookston (1979)	*Triticum aestivum* (spring wheat, 2 vars.)	5–17 days (after anthesis)	Kernel W	$\bar{G}$	Analysis unduly restricted by function
Sims (1979a)	*Brassica* sp. (oilseed rape)	8, 20 days (after seed-set)	$[H_2O]$	$-\bar{G}$	Moisture loss during ripening
Sims (1979b)	*Brassica* sp. (oilseed rape)	Various periods of 14 days after cutting	Seed W ha^{-1} $[H_2O]$	$-\bar{G}$	Moisture loss during ripening, also dry wt. loss
Vil'yams *et al.* (1979)	*Beta vulgaris* (sugar beet)	35–105 days (p)	Concentrations of N, P, K, Ca, Mg, CH_2O		
Asimi, Gianninazzi-Pearson and Gianninazzi (1980)	*Glycine max* (soybean)	45–115 days (g)	N_2 fixation	$\bar{G}$	
Erdős (1980)	*Zea mays* (maize)	100 years	Grain W		Long-term trends
Herrera and Ramos (1980)	*Cynodon dactylon* (Bermuda grass)	7–84 days (g)	shoot [P], [K]	$\bar{G}$	
Holliday and Putwain (1980)	*Senecio vulgaris* (groundsel)	12 years (e)	% mortality	$-\bar{G}$	Simazine effects
Linhart and Wheelan (1980)	*Acer pseudoplatanus* (sycamore)	23 years (p)	Basal stem diam.		Non-Gaussian replicates
Major (1980)	*Zea mays* (maize)	30–95 days (after anthesis)	$[H_2O]$	$-\bar{G}$	Moisture loss during ripening

Table 5.1 (*Continued*)

Author(s)	Species	Time interval(s) (see footnote)	Primary data	Derived data	Comments
Marrs, Roberts and Bradshaw (1980)	Community on reclaimed china clay waste	90 months (e)	Community S_W	$\bar{G}$	
Meyer (1980)	3 deciduous trees	300 days	Litter P	$-\bar{G}$	Absolute rates of decomposition
Ormrod *et al.* (1980)	*Calendula officinalis* (marigold)	7–14 days (g)	Cotyledon length		
Pulli (1980a)	*Phleum pratense* (timothy)	10–35 days (g)	Length, H, W	$\bar{G}$	T adjusted to 'growing days'
Simpson and Gumbs (1980)	*Saccharum officinarum* (sugar cane)	13–22 weeks (p)	Total H	$\bar{R}$	
Smyth and Dugger (1980)	*Cylindrotheca fusiformis* (a diatom)	30 min (e)	^{86}Rb counts	$\bar{G}$	Influx/efflux rates for ^{86}Rb
Logarithmically transformed primary data					
Roach (1926)	*Scenedesmus costulatus* (a soil alga)	up to 15 days (e)	V, N	$\bar{R}$	
Gregory (1928b)	*Cucumis sativus* (cucumber)	16 days (g)	L_A	$\bar{R}$	
Roach (1928)	*Scenedesmus costulatus*	10 days (e)	V	$\bar{R}$	
Ashby (1929)	*Lemma minor* (duckweed)	10 days (e)	W	$\bar{R}$	
Ashby (1930)	*Zea mays* (maize, two parents and the F_1 hybrid)	71 days (g)	W	$\bar{R}$	
Ashby (1932)	*Zea mays* (maize)	7–50 days (g)	W, FW	$\bar{R}$	
Heath (1932)	*Gossypium hirsutum* (cotton)	20–70 days (g)	H	$\bar{R}$	Full statistics included
Afzal and Iyer (1934)	*Gossypium hirsutum* (cotton)	20–70 days (g)	H	$\bar{R}$	Full statistics included

(*Continued*)

Table 5.1 (*Continued*)

Author(s)	Species	Time interval(s) (see footnote)	Primary data	Derived data	Comments
Hicks and Ashby (1934)	*Lemna minor* (duckweed)	10 days (e)	N	$\bar{R}$	Many temperature treatments
Portsmouth (1937)	*Cucumis sativus* (cucumber)	2–22 days (g)	L, L_A FW, W	$\bar{R}$	
Ashby (1937) (data from Jones, 1918)	*Zea mays* (maize)	40–90 days (p)	H	$\bar{R}$	
Heath (1937b)	*Gossypium hirsutum* (cotton)	8–78 days (g)	H, W	$\bar{R}$	
Bond (1945)	*Camellia thea* (tea)	35 days (e)	Leaf initials, length	$\bar{R}$	
Muramoto, Hesketh and El-Sharkawy (1965)	*Gossypium* spp. (cotton)	1–8 weeks (g)	L_A	$\bar{R}$	
Troughton (1965)	*Lolium perenne* (perennial ryegrass)	1–5 weeks (e)	V, N	$\bar{R}$s	V by the Archimedes method
Lupton, Ali and Subramaniam (1967)	*Triticum aestivum* (wheat)	17 weeks (e)	L_A, tiller N	$\bar{R}$, E	Orthogonal polynomials, E is only $1/L_A \cdot dL_W/dT$
Troughton (1967)	*Lolium perenne* (perennial ryegrass)	Up to 35 days (e)	V, N	$\bar{R}$s	V by the Archimedes method
Duncan and Hesketh (1968)	*Zea mays* (maize, 22 vars.), *Euchlaena mexicana* (teosinte)	15–50 days (e)	L_A	$\bar{R}$	
Troughton (1968)	*Lolium perenne* (perennial ryegrass)	35 days (e)	Length, N, W	$\bar{R}$s	
Brewster and Tinker (1970)	*Allium porrum* (leek)	49–84 days (p)	H_2O, R_L	Mean water inflow	A 'type (iii)' derivate
Scott (1970)	12 N.Z. and introduced grasses	'to 0.2–0.8 g W'	W	$\bar{R}$	
Williams and Bouma (1970)	*Trifolium subterraneum* (subterranean clover)	2–27 days (p)	L length, L_A, L_V	$\bar{R}$	See Fig. 5.6

Table 5.1 (*Continued*)

Author(s)	Species	Time interval(s) (see footnote)	Primary data	Derived data	Comments
Silsbury (1971)	*Lolium perenne* (perennial ryegrass)	32 days (p)	W, L_A	$\bar{R}$s, F, E	
Troughton (1971)	*Lolium perenne* (perennial ryegrass)	*c.* 8 weeks (12 expts.)	Tiller N, S_W, S_V	$\bar{R}$s	
Bean and Yok-Hwa (1972)	*Lolium perenne* (perennial ryegrass, 7 vars.), *L. multiflorum* (Italian ryegrass, 5 vars.)	25 days (e)	W, L_A	$\bar{R}$s, F, E	
Leshem *et al.* (1972)	*Dactylis glomerata* (cocksfoot)	up to 3 hours (e)	H_2O content	$-\bar{R}$	Water loss
Macdowall (1972a, b, c)	*Triticum aestivum* (Marquis spring wheat)	7–28 days (g)	Ws, L_A, FWs, chlorophyll	$\bar{R}$s	NO_3^-, light and CO_2 levels varied
Hunt (1973)	*Rumex acetosa* (sorrel)	14–42 days (g)	P, R_W	A_P	
Macdowall (1973a, b)	*Triticum aestivum* (Marquis spring wheat)	7–28 (–56) days (g)	Ws, L_A, FWs	$\bar{R}$s	Temperature and growth of components involved
Macdowall (1974)	*Triticum aestivum* (spring wheat, 6 vars.)	7–42 (–68) days (g)	W	$\bar{R}$	
Grime and Hunt (1975)	132 spp., mainly British natives	14–35 days (g)	W	$\bar{R}$	
Murata (1975a)	*Oryza sativa* (rice)	180 days (e)	[N]	$-\bar{R}$	Decay rates for N content
Murata (1975b)	*Zea mays* (maize), *Glycine max* (soybean)	140 days (p)	[N]	$-\bar{R}$	Decay rates for N content
Hodgkinson and Quinn (1976)	*Danthonia caespitosa* (tussock grass, 5 vars.)	18–53 days (p)	W	$\bar{R}$	$\bar{R} \propto$ latitude

(*Continued*)

Table 5.1 (*Continued*)

Author(s)	Species	Time interval(s) (see footnote)	Primary data	Derived data	Comments
Raper, Weeks and Wann (1976)	*Nicotiana tabacum* (tobacco)	42 days (p)	*W*s, N, L_A CH_2O	$\bar{R}$s	Very many combinations of variable × component
Barrow (1977)	6 tree spp.	26–133 days (p)	W, P, L_W	$\bar{R}$, E	
Gordon, Balaam and and Derera (1977)	*Triticum aestivum* (wheat, 4 vars.)	70 days (after anthesis)	grain W	$\bar{R}$	Logits also employed
Potter and Jones (1977)	9 crops and 'weeds'	Up to 5 g W	L_W, L_A, W	$\bar{R}$s, ratios, E	
Raper (1977)	*Nicotiana tabacum* (tobacco)	7–32 days (p)	*W*s and nutrient contents for many components	$\bar{R}$s	
Raper *et al.* (1977)	*Gossypium hirsutum* (cotton), *Glycine max* (soybean)	22–56 days or 16–43 days (p)	*W*s, Ns, L_A	$\bar{R}$s	
Farrar (1978)	*Hypogymnia physodes* (a lichen)	50 hours (e)	W, CO_2 flux	$\bar{R}$	
Hüsken, Steudle and and Zimmermann (1978)	*Capsicum annuum* (green pepper)	13 min (e)	H_2O flux	$-\bar{R}$	See Fig. 5.4
Jones *et al.* (1978)	*Glycine max* (soybean)	12 hours (e)	^{45}Ca	$-\bar{R}$	Efflux rates
Tullberg and Angus (1978)	*Medicago sativa* (lucerne)	30 (or 60) hours (e)	Shoot H_2O	$-\bar{R}$	Drying rate
Busey and Myers (1979)	8 subtropical turf-grasses	46 days (p)	$_2FW/_1FW$	$\bar{R}$	Primary datum is 'growth quotient'
Elias and Chadwick (1979)	40 grass and legume cultivars	14–35 days (g)	*W*s	$\bar{R}$s	
Gordon (1979)	*Triticum aestivum* (wheat)	63 days (after anthesis)	Germinative α-amylase	$\bar{R}$s	Logits also employed

Table 5.1 (*Continued*)

Author(s)	Species	Time interval(s) (see footnote)	Primary data	Derived data	Comments
Rufty, Miner and Raper (1979)	*Nicotiana tabacum* (tobacco)	42 days (e)	*W*s, Mn	$\bar{R}$s	
Farrar (1980)	*Hordeum distichum* (barley)	22 days (g)	^{14}C content	$\bar{R}$	
Burdon and Harper (1980)	*Trifolium repens* (48 clones)	28 days (e)	*W*	$\bar{R}$	See section 2.2.2
Caloin, El Khodre and Airy (1980)	*Dactylis glomerata* (cocksfoot)	28–*c.* 42 days (p)	W, R_W, S_W	$\bar{R}$	Initial slope of curvilinear trends
Gloaguen and Touffet (1980)	Woody spp. from Britanny	1800 days (e)	Litter *W*	$-\bar{R}$	Rates of decay
Hedley and Ambrose (1980)	*Pisum sativum* (garden pea, 6 vars.)	15 days (after anthesis)	Seed *FW*, *W*	$\bar{R}$	
Meyer (1980)	3 N. American trees	600 days (e)	Litter *W*	$\bar{R}$	Rates of decay
Rumberg, Ludwick and Siemer (1980)	*Bromus inermis* (bromegrass), *Phleum pratense* (timothy)	10 weeks (e)	*W*	$\bar{R}$	
Townsend and McRae (1980)	*Zea mays* (maize)	48–97 days (from seeding)	NO_3–N	$-\bar{R}$	Rate of decline of of NO_3–N in seed
Wallén (1980)	*Calluna vulgaris* (heather)	2–12 years	Tiller *W*	$\bar{R}$	
van de Dijk (1980)	*Hypochaeris radicata* (cat's ear)	16–29 days (p)	*W*	$\bar{R}$	

Notes: One or the other of the scientific and common names of the species is that given by the original author(s): time intervals show the *maximum* period over which the analysis was conducted; this was measured: (e) from the beginning of measurements made on material already established, (g) from germination or emergence, (p) from planting; frequency of sampling is not taken into account; for symbols for primary data see Table 2.4, also used are conventional chemical symbols and: *V* volume, *N* number, *H* height, ΔW increment in *W*, L (as a subscript) length; some symbols appear as plurals, e.g. *W*s, dry weights. All of these notes also apply to succeeding Tables of this type.

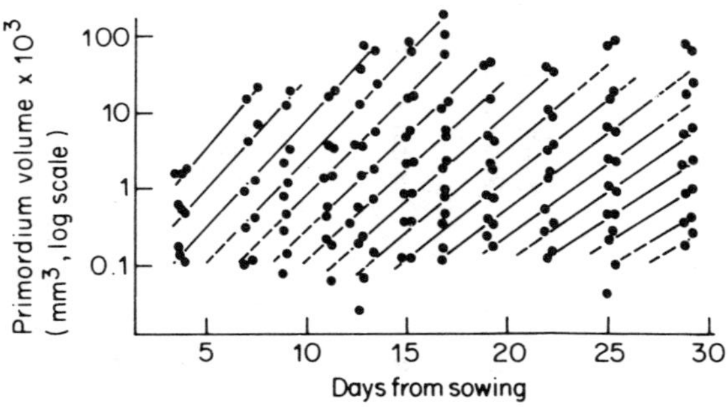

Fig. 5.6 First-order polynomial exponential fits to volumes of corpus tissue associated with successive leaf primordia in *Trifolium subterraneum* (data and analyses from Williams and Bouma, 1970).

has frequently been attempted, often with justification. However, we see a predominance of lengths, numbers, volumes and concentrations among the primary data and conclude that these variates, rather than dry weight itself, most often fulfil the special requirements that need to be satisfied before proceeding in this way (section 4.7.2). In one instance (MacColl, 1977), a first-order polynomial forms a vital part of an entirely valid analysis proceeding right through to **F**, **E** and **L** (Fig. 5.9).

The major use of the function, however, has been in the calculation of $\bar{\mathbf{R}}$ (or $-\bar{\mathbf{R}}$, the relative decay rate) from data transformed to natural logarithms. This powerful comparative tool has seen very many large-scale applications (e.g. Duncan and Hesketh, 1968; Scott, 1970; Macdowall, 1972a *et seq.*; Grime and Hunt, 1975; Elias and Chadwick, 1979; Burdon and Harper, 1980), but this section concludes with one illustrated example only, the remarkable display of log-linear pyrotechnics from Williams and Bouma (1970) presented in Fig. 5.6. Further examples appeared in Figs. 4.7 and 4.8b.

5.3 Second-order polynomial

5.3.1 The function and its properties

This function, also known as the quadratic or parabolic, is

$$W\,(\text{or}\ \log_e W) = a + b_1 T + b_2 T^2 \tag{5.10}$$

with the same provisos regarding the units of parameter a (section 5.2.1). From this,

$$\frac{dW}{dT}\left(\text{or}\ \frac{1}{W}\cdot\frac{dW}{dT}\right) = b_1 + 2b_2 T \tag{5.11}$$

As before, the untransformed function leads directly to absolute growth rate and the transformed function to relative growth rate. From equation 5.11 we see that these derivates both change linearly with time, so the function can provide for the simplest possible expression of the clear departure from linear or log-linear progressions of W on time which ultimately occurs during the growth of all multicellular organisms. For this reason, some workers (e.g. Eagles, 1969; Buttery and Buzzell, 1974; Hurd, 1977) prefer the function to the exclusion of others, provided that very advanced stages of growth are not involved. Indeed, used in this way the function has much to recommend it as the plainest and simplest of 'growth curves'.

The curvature of quadratics is always uniform and progressive. Parameter a represents, as in all polynomials, the size at $T = 0$, parameter b_1 represents **G** or **R** at $T = 0$ and parameter b_2 reflects the degree of curvature, or rate of change, in progressions of **G** or **R** (see section 3.4.8). In Fig. 5.7 six variants of the function are displayed as curvilinear progressions of $\log_e W$ on T, with the resulting linear progressions of slope. By adjusting the sign and

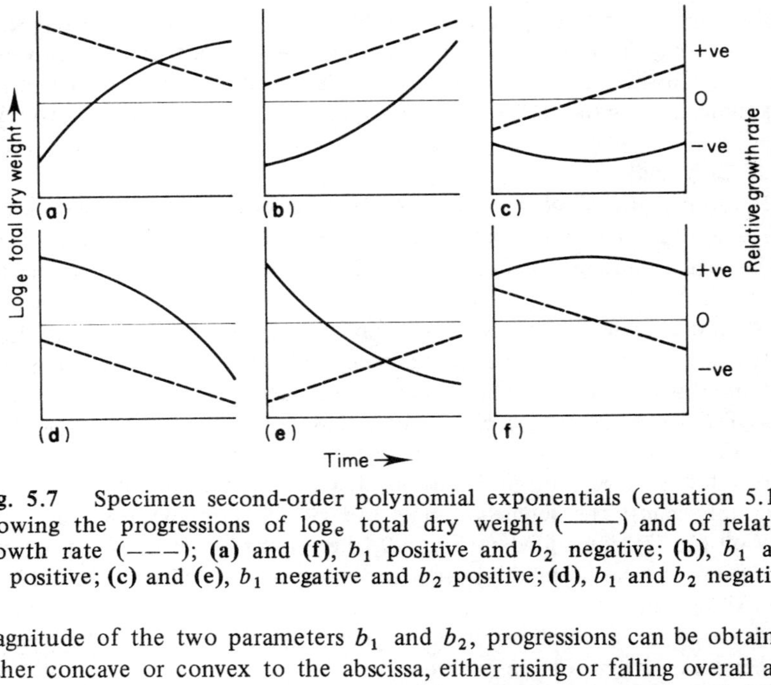

Fig. 5.7 Specimen second-order polynomial exponentials (equation 5.10) showing the progressions of $\log_e$ total dry weight (——) and of relative growth rate (– – –); **(a)** and **(f)**, b_1 positive and b_2 negative; **(b)**, b_1 and b_2 positive; **(c)** and **(e)**, b_1 negative and b_2 positive; **(d)**, b_1 and b_2 negative.

magnitude of the two parameters b_1 and b_2, progressions can be obtained either concave or convex to the abscissa, either rising or falling overall and either with or without peaks or troughs (see the legend to Fig. 5.7). What cannot be accommodated are inflections in the fitted curve and the consequent non-linearity in the progressions of **G** or **R**.

Kreusler's data for maize (Table 1.1), when fitted by the second-order polynomial exponential (Fig. 5.8), called for a very small (negative) value of b_2 and hence an extremely slight curvature in the fitted function, the

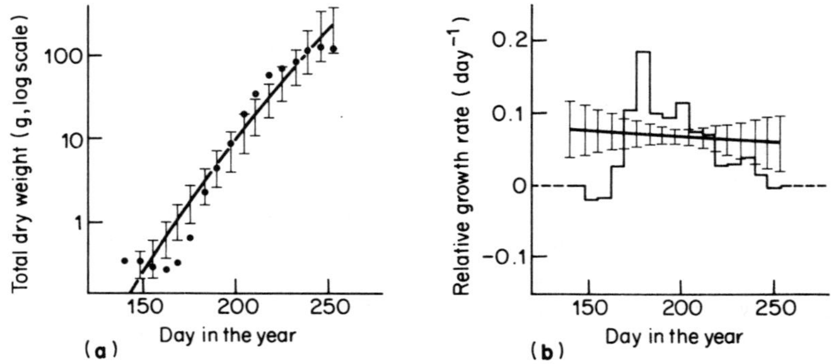

Fig. 5.8 Kreusler's data for maize (Table 1.1) fitted by a second-order polynomial exponential; **(a)** observed values of total dry weight per plant, with the fitted curve and 95% limits; **(b)** fitted values of relative growth rate, with 95% limits, shown against the harvest-interval means.

shallow decline in **R** ranging from 0.078 to 0.059 day^{-1}. This analysis, of course, is only fractionally more realistic than the log-linear analysis shown in Fig. 5.2. We should notice, too, how the limits of **R** have widened in the meantime, making the addition of the b_2 parameter a retrograde step overall.

5.3.2 Case studies

Work involving fitted second-order polynomial growth curves is listed in Table 5.2. Again, the selection avoids examples involving more than two orders of polynomial simultaneously (section 5.6), or the segmented approach (see section 7.2).

When dealing with untransformed primary data the same remarks apply as in the case of the first-order polynomial (section 5.2.2). Indeed, it seems that untransformed applications may outnumber the transformed, though many fewer proceed to any derived data than is the case with the first-order curve (Table 5.1): this is clearly a wasted opportunity. There are, too, more than a few instances of problems involving heteroscedasticity of variance here, for against the serenely simple and virtually trouble-free analysis of MacColl (1977), illustrated in Fig. 5.9, there must be set the unnecessarily (but properly and inevitably) truncated analyses of Allison (1969, 1971) and Allison and Watson (1966) (see Fig. 4.9), who found the alliance of quadratic curve and untransformed primary data no match for the realities of their extensive runs of observations.

Using transformed data, workers have been rather more adventurous in obtaining derivates from their curves, two recent and perfectly satisfactory examples being those of Farrar (1980) and Freyman (1980) for plants grown as individuals and in closed stands respectively, while, using the function for purely representational purposes, what could be better than Idris and Milthorpe's (1966) neat family of curves (Fig. 5.10)?

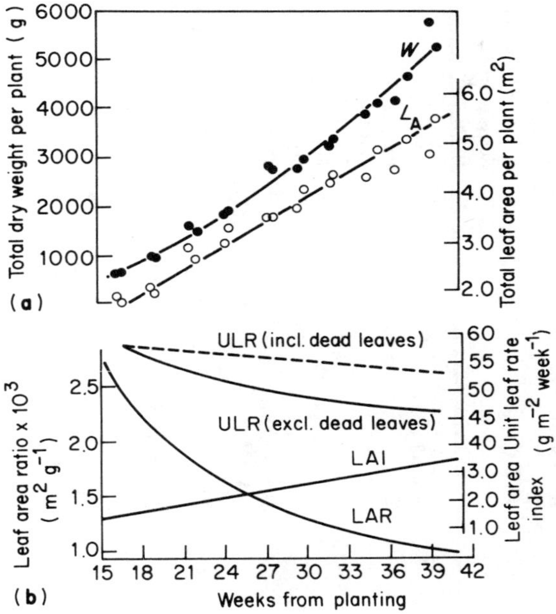

Fig. 5.9 (a) Second-order polynomial fits to total dry weight and total leaf area (both per plant) in sugar cane; (b) derivates from the curves shown in (a) (data and analyses from MacColl, 1977).

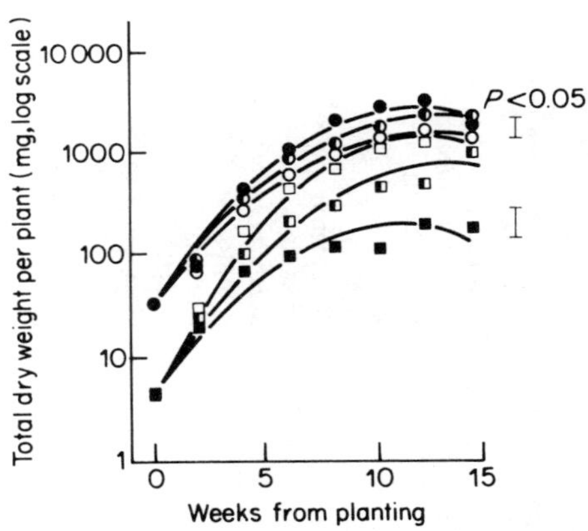

Fig. 5.10 Families of second-order polynomial exponentials fitted to total dry weight per plant in barley (circles) when grown in monoculture (○), in excess (◐), and subordinate in numbers to charlock (●); and in charlock (squares) when grown in monoculture (□), in excess (◧), and subordinate in numbers to barley (■) (data and analyses from Idris and Milthorpe, 1966).

Table 5.2 Applications of the second-order polynomial (untransformed primary data) and second-order polynomial exponential (logarithmically-transformed primary data).

Author(s)	Species	Time interval(s)	Primary data	Derived data	Comments
Untransformed primary data					
Kershaw (1962)	*Carex bigelowii* (stiff sedge)	2–27 years	L_L, L_{width}		
Vernon and Allison (1963)	*Zea mays* (maize)	70 days (after anthesis)	W, L_A	$\mathbf{E}_A$	The first 'type (iii)' derivate from fitted curves
Allison and Watson (1966)	*Zea mays* (maize)	56 days (e)	**D**, **E**, also **E** = $f(T)$		
Humphries (1968)	*Phaseolus vulgaris* (dwarf bean)	28 days (e)	L_A		Parameter b, omitted
Allison (1969)	*Zea mays* (maize)	10–18(–22) weeks (p)	L_A, W	**C**, **L**	Analysis truncated by heteroscedasticity
Hearn (1969a)	*Gossypium* spp. (cotton vars.)	15–140 days (p)	Mainstem node N		Day degrees also used
Moorby (1970)	*Solanum tuberosum* (potato, var. Arran pilot)	91 days (p)	L_A, Ws	**E**	
Allison (1971)	*Zea mays* (maize)	Various	W, L_A	**R**, **F**, **E**	Analyses truncated by heteroscedasticity
Hearn (1972a, b)	*Gossypium* sp.	80–130 days (p)	W, flower N, **L**	**R**, **C**, **E**	
Sobulo (1972)	*Dioscorea rotundata* (white yam)	169 days (e)	W, L_W	**L**, **E**	Derivates on basis of L_W only
Thornley and Hesketh (1972)	*Gossypium* sp.	40 days (e)	Specific respiration rate in cotton balls	Specific maintenance respiration rate	
Matthews (1973)	*Pisum sativum* (pea, var. Kelvedon wonder)	20–70 days (after fertilization)	Many seed variates		
Scott, *et al.* (1973)	*Beta vulgaris* (sugar beet)	125–245 days (p)	Sugar yield		Various time axes used

Table 5.2 (*Continued*)

Author(s)	Species	Time interval(s)	Primary data	Derived data	Comments
Murata (1975b)	*Glycine max* (soybean)	20–90 days (p)	[N]		
Walton (1976)	*Acaena magellanica* (dwarf shrub, Rosaceae)	3–7 years (e)	S_W, S_N		
MacColl (1977)	*Saccharum* sp. (sugar cane, 12 vars.)	15–42 weeks (p)	W	**F, E, L**	See Fig. 5.9
Barber (1978)	*Glycine max* (soybean)	30–120 days (g)	R_L		
Graves (1978)	*Chrysanthemum morifolium* (chrysanthemum)	10–100 days (e)	Cu content		
Rüegg and Alston (1978)	*Medicago truncatula* (barrel medic)	49–134 days (g)	C_2H_2 reduction, N_2 accumulation		
Yukimura and Kanda (1978)	*Avena sativa* (oat)	10–38 days (p)	W, L_A, H	**G**	Authors' 'RGR' is **G**
Christie (1979)	Semi-arid grassland community	84 days (p)	R_W, N, P, litter W		
English *et al.* (1979)	*Helianthus annuus* (sunflower)	12 hours (diurnal trends)	L water pot.		
Sims (1979b)	*Brassica* sp. (oilseed rape)	35 days (e)	Seed W ha^{-1}, [H_2O], oil yield ha^{-1}		
Vil'yams *et al.* (1979)	*Beta vulgaris* (sugar beet)	35–105 days	W and concentrations of N, P, K, Ca, Mg, CH_2O		
Asimi, Gianninazzi-Pearson and Gianninazzi (1980)	*Glycine max* (soybean)	45–115 days (g)	N_2 fixation		
Gloaguen and Touffet (1980), Gloaguen, Touffet and Forgeard (1980)	Woody spp. from Brittany	1800 days (e)	Litter W	$-\bar{\mathbf{R}}$	Times to 25% or 50% of losses predicted

(*Continued*)

Table 5.2 (*Continued*)

Author(s)	Species	Time interval(s)	Primary data	Derived data	Comments
Herrera and Ramos (1980)	*Cynodon dactylon* (Bermuda grass)	7–84 days (g)	Shoot Ca, Mg		
Kappelman (1980)	*Gossypium* spp. (cotton)	9 years	'Relative mean wilting percentage'		*Fusarium* epidemology
Phatak *et al.* (1980)	*Brassica campestris* (turnip greens)	20–41 days (p)	L_N, L_L, L_{FW}, L_W		Many soil and tillage comparisons
Pulli (1980b)	*Festuca pratensis* (meadow fescue)	40–95 days (g)	W		
Wallén (1980)	*Calluna vulgaris* (heather)	2–12 years (e)	Stem diam.		
Logarithmically-transformed primary data					
Heath (1937a)	*Gossypium hirsutum* (cotton)	8–78 days (g)	W, L_W	**R**	See section 3.4.8
Muramoto, Hesketh and El-Sharkawy (1965)	*Gossypium* spp. (cotton)	7–56 days (g)	W		
Idris and Milthorpe (1966)	*Hordeum vulgare* (barley)	14 weeks (p)	W		See Fig. 5.10
Bremner, El Saeed and Scott (1967)	*Beta vulgaris* (sugar beet)	56 days (p)	W	**C**	
Lupton, Ali and Subramaniam (1967)	*Triticum aestivum* (wheat, 5 vars.)	17 weeks (e)	S_W	**E, R**	Orthogonal polynomials, E on dry wt bases
Buttery (1969)	*Glycine max* (soybean)	30–120 days (p)	W, L_A	**Rs, E, C, L**	Full statistics given
Eagles (1969)	*Dactylis glomerata* (cocksfoot, 2 vars.)	7–42 days (e)	W	**R**	
Hearn (1969b)	*Gossypium* spp. (cotton vars.)	15–140 days (p)	W	**C**	

Table 5.2 (*Continued*)

Author(s)	Species	Time interval(s)	Primary data	Derived data	Comments
Silsbury (1969)	*Lolium perenne* (perennial ryegrass), *L. rigidum* (an annual ryegrass)	32 days (p)	W	**R**	
Eagles (1971)	*Dactylis glomerata* (cocksfoot, 2 vars.)	7–42 days (e)	W, L_A	**Rs, F, E**	
Monyo and Whittington (1970, 1971)	*Triticum aestivum* (vars. of wheat)	12 weeks (p)	Ws, L_A	**R, dR/dT, E**	Orthogonal polynomials employed
Silsbury (1971)	*Lolium perenne* (perennial ryegrass)	32 days (p)	W, L_A	**Rs, E**	
Bean and Yok-Hwa (1972)	*Lolium perenne* (perennial ryegrass, 7 vars.), *L. multiflorum* (Italian ryegrass, 5 vars.)	25 days (e)	W, L_A	**Rs, F, E**	
Buttery and Buzzell (1974)	*Glycine max* (soybean)	21–84 days (p)	W, L_A	**E**	
Dudney (1974)	*Malus* sp. (apple, 3 vars.)	9 years (e)	W, W_P	**Gs, Rs, H**	See section 2.2.7
Hurd (1977)	*Lycopersicon esculentum* (tomato)	10–48 days (p)	W, L_A	**Rs, F, E**	See section 5.6
Stanhill (1977)	*Daucus carota* (carrot)	150 days (g)	R_{FW}, L_{FW}		
Clarke and Simpson (1978)	*Brassica napus* (rape)	60 days (e)	W	Classical derivates	
Doley (1978)	*Eucalyptus grandis* (a timber eucalypt)	7–49 (days (g)	Ws, L_A	**Rs**, SLA, ΔW	
Neales and Nicholls (1978)	*Triticum aestivum* (wheat)	10–26 days (g)	W, L_A	**Rs, F**, SLA, LWR, **E**	

(*Continued*)

Table 5.2 (*Continued*)

Author(s)	Species interval(s)	Time interval(s)	Primary data	Derived data	Comments
Sivakumar and Shaw (1978)	*Glycine max* (soybean)	35–112 days (p)	W, L_A	**Rs, E, C**	
Bhat, Brereton and Nye (1979a, b)	*Brassica napus* (rape)	12(24) days (e)	Ws, FWs, L_A R_A, [N]	**Rs, A, E, F**	Mechanistic model attempted
Bhat, Nye and Brereton (1979)	*Brassica napus* (rape)	24 days (e)	Ws, FWs, L_A,	**Rs, A, E, F**	Mechanistic model attempted
Fribourg *et al.* (1979)	*Dactylis glomerata* (cock's foot) *Trifolium repens* (white clover)	140 days (p)	Digestibility		
Friedrich and Schrader (1979)	*Zea mays* (maize)	7 weeks (after flowering)	S, N content		With full statistics
Gordon (1979)	*Triticum aestivum* (wheat)	63 days (after anthesis)	Germinative α-amylase, flavanol percentage		Also quadratic fits to logit % dormancy
Ho and Shaw (1979)	*Lycopersicon esculentum* (tomato)	10–30 days (e)	L_A, L_{FW}, L_W L_{carbon}, $L_{minerals}$	**Rs**, accumulation rates, mineral concentrations, carbon balance sheet	Seventh leaf only
Farrar (1980)	*Hordeum distichum* (barley)	30 days (g)	W, L_A	**R, F, E**	
Freyman (1980)	*Hordeum* sp. (barley)	80 days (p)	W, L_A	**Rs, E, C, L**	With full statistics
Ormrod *et al.* (1980)	*Calendula officinalis* (marigold)	7–14 days (g)	W, cotyledon length		Includes inter-lab. comparisons
Smith and Rogan (1980)	*Agropyron repens* (couch grass)	45–105 days (g)	W		
Townsend and McRae (1980)	*Zea mays* (maize)	48–97 days (from seeding)	NO_3–N		

For footnotes, see Table 5.1.

5.4 Third-order polynomial

5.4.1 The function and its properties

This function, also known as the cubic, is

$$W\,(\text{or}\ \log_e W) = a + b_1T + b_2T^2 + b_3T^3 \tag{5.12}$$

from which

$$\frac{\mathrm{d}W}{\mathrm{d}T}\left(\text{or}\ \frac{1}{W}\cdot\frac{\mathrm{d}W}{\mathrm{d}T}\right) = b_1 + 2b_2T + 3b_3T^2 \tag{5.13}$$

The slope of the function exhibits a progression on time which is a second-order polynomial and the function itself is always either continuously curving in one direction, or S-shaped. Coefficient ***a*** behaves as usual, but it is impossible to ascribe biological significance to the b coefficients in all of their eight permutations of sign. Three common instances are examplified in Fig. 5.11

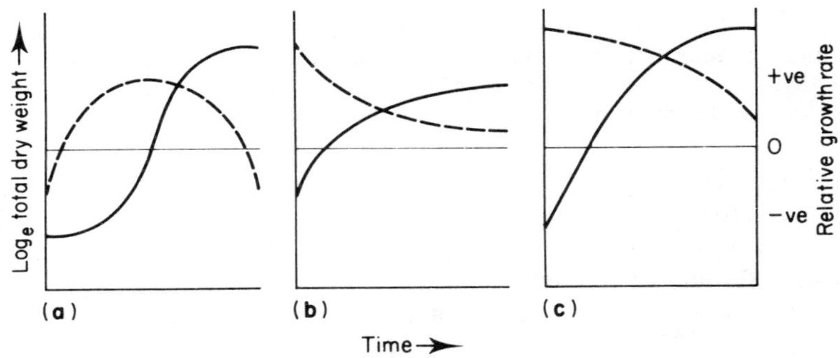

Fig. 5.11 Specimen third-order polynomial exponentials (equation 5.12) showing the progressions of total dry weight (——) and of relative growth rate (– – –); **(a)**, b_1 negative, b_2 positive, b_3 negative; **(b)**, b_1 positive, b_2 negative, b_3 positive; **(c)**, b_1 positive, b_2 negative, b_3 negative.

and even of these, many other variants differing in overall direction are possible. From the first $(-b_1 + b_2 - b_3)$, a dome-shaped progression of slope upon time results; from the second $(+b_1 - b_2 + b_3)$, arises a gradually diminishing decline in slope; while in the third $(+b_1 - b_2 - b_3)$, a similar decline accelerates. In this last case, coefficient b_1 represents the initial slope, b_2 half of its initial rate of decline and b_3 one sixth of the acceleration in that decline. Such conceptual complexity means that most often the third-order polynomial is used simply as an empirical approximating function to fit progressions exhibiting one simple inflection, or one simple direction of curvature which changes in curvilinearity as the data progress. Always, the form of the curve is determined by the signs of its coefficients and its degree of curvilinearity by their magnitudes.

Kreusler's data for maize (Table 1.1), when fitted by the third-order polynomial exponential (Fig. 5.12), show an almost symmetrical S-shaped progression of the fitted values on time (cf. Fig. 5.11). Superficially at least,

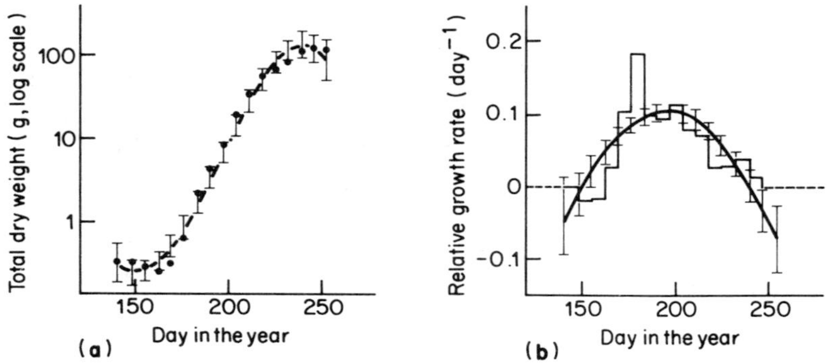

Fig. 5.12 Kreusler's data for maize (Table 1.1) fitted by a third-order polynomial exponential; **(a)** observed values of total dry weight per plant, with the fitted curve and 95% limits; **(b)** fitted values of relative growth rate, with 95% limits, shown against the harvest-interval means.

this brings the first sign in our series (Figs 5.2, 5.8) of a satisfactory analysis. But, the progression of **R** on time (Fig. 5.12b), though resembling that of $\bar{\mathbf{R}}$ inasmuch as a period of high, positive **R** intervenes between periods of negative **R**, smothers several important points of detail. For example, the magnitude of the peak is probably underestimated and the phase of negative $\bar{\mathbf{R}}$ is most marked at the beginning of growth, not at the end. For these reasons, a sharper-peaked distribution of **R** skewed towards the earlier stages could be considered desirable, but the function, alas, is not equipped to provide this (Hunt and Parsons, 1977, discuss this analysis further).

5.4.2 Case studies

Work involving fitted third-order polynomials is listed in Table 3.3, with the usual provisos.

Again, analyses involving untransformed primary data, often with several derivates, are far from rare, even though the obvious need of associating third-order polynomials with extended series of observations inevitably brings danger from value-dependent variability, such as that experienced by Goodman (1968) and Allison (1971). Noteworthy among the collection listed in the first part of Table 5.3 are Graves's evaluation of specific absorption rate for copper (1978), Huzulák and Matejka's neat use of the cubic as a 'pseudo-sine function' to define diurnal trends in xylem pressure potential (1980), and Wilson, Clowes and Allison's inter-varietal comparisons of maize (1973) which, curiously, despite satisfactory fitted curves for W (Fig. 5.13), involved classically-based analyses of change in L_A.

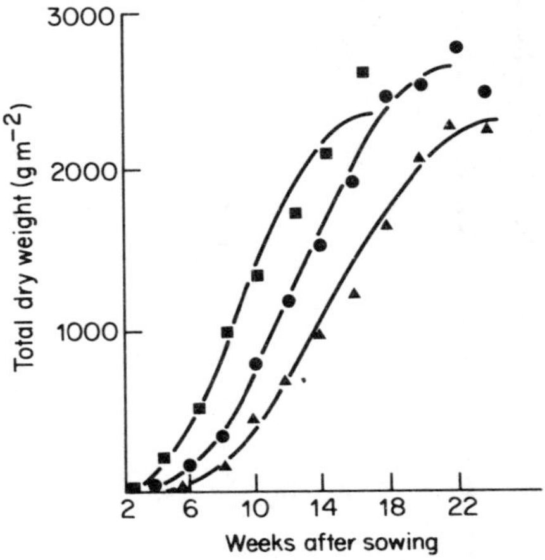

Fig. 5.13 Third-order polynomials fitted to total dry weight (m^{-2}) in maize grown at three altitudes in Rhodesia; ■, Chiredzi (420 m); ●, Henderson (1260 m); ▲, Grasslands (1620 m). Short vertical lines indicate times of flowering (data and analyses from Wilson, Clowes and Allison, 1973).

In the case of the third-order polynomial exponential, the picture is dominated by the work of A. P. Hughes, his associates and followers. The superficial resemblance of the function, in one of its guises, to the decline, rise and plateau commonly seen in full series of data on the growth from seed of an annual plant (see Fig. 1.1b) led Hughes to recommend the function for widespread use on semi-biological grounds (p. 73). When dealing solely with representational fits to primary data such usage has often been entirely successful, but a difficulty frequently arises when estimating further derivates, particularly those of 'type (iii)' (see section 2.1). We saw in section 5.2.1 that even first-order fits to primary data allowed relatively free progression in, say, ULR. When these fits rise to the third-order, this freedom can become positively harmful, with terminal elements of the progressions of ULR threshing wilfully about in response to any chance unrepresentativeness that may be present in the terminal points or arrays in the series of primary data. Figure 5.14, from Hughes and Freeman's own exposition of the method (1967), shows this well: despite superficially satisfactory fits to the primary data (e.g. for W in part (a), L_A was very similarly treated), an entirely spurious late upturn in ULR is the result (part (b)). Though discounted by its limits, the upturn in this progression disguises reality to a considerable degree (Evans, 1972, p. 341; Hunt and Parsons, 1974).

Table 5.3 Applications of the third-order polynomial (untransformed primary data) and third-order polynomial exponential (logarithmically-transformed data).

Author(s)	Species	Time interval(s)	Primary data	Derived data	Comments
Untransformed primary data					
Goodman (1968)	*Beta vulgaris* (sugar beet)	16 weeks (p)	*W*, **L**	**E**	Heteroscedasticity, classical derivates also
Allison (1971)	*Zea mays* (maize)	7–23 weeks (p)	*W*	**R, E, F,**	Quadratic L_A, heteroscedasticity
Kornher (1971)	*Festuca pratensis* (meadow fescue), *Phleum pratense* (Timothy)	30–90 days (p)	*W*, **L**	**C, R, E**	
Wilson, Clowes and Allison (1973)	*Zea mays* (maize)	2–23 weeks (p)	*W*	**C**	Classical analysis of L_A, see Fig. 5.13
Arnott (1975)	*Lolium perenne* (perennial ryegrass)	3–10 days	*W*s		
Graves (1978)	*Chrysanthemum morifolium*	10–100 days (e)	Cu content	**R**, $\mathbf{A}_{\mathbf{Cu}}$	
Baker (1979)	3 Ghanaian cereals	168 days (p)	*H*		
Cock *et al.* (1979)	*Manihot esculenta* (cassava)	5–50 weeks (p)	Leaf *N*	**R**	
Friedrich, Schrader and Nordheim (1979)	*Zea mays* (maize)	16 weeks (e)	NO_3-N		
Lawn (1979a, b)	4 *Vigna* spp. (cow peas)	112 days (p)	*W*, seed *W*	Seed weight ratio	
Huzulák and Matejka (1980)	3 deciduous tree sp.	24 hours (diurnal trend)	Xylem pressure potential		
Watanabe and Takahashi (1979)	*Dactylis glomerata* (cock's foot)	8 weeks (e)	*W*, L_A	**C, F, L, E**	
Migus and Hunt (1980)	*Triticum aestivum* (wheat, 2 vars.)	100–150 days (p)	CO_2 exchange rate [N], transpiration rate		

Table 5.3 (*Continued*)

Author(s)	Species	Time interval(s)	Primary data	Derived data	Comments
Pulli (1980b)	*Festuca pratensis* (meadow fescue)	40–95 days (g)	**L**, *H*, *W* protein cellulase		*T* also adjusted to 'growing days'
Logarithmically transformed primary data					
Williams (1964)	*Triticum aestivum* (wheat)	25 weeks	S_W	**R**	Compared with logistic and segmented fits
Hughes and Freeman (1967)	*Callistephus chinensis* (China aster)	49–88 days (p)	W, L_A	**R, F, E**	See Fig. 5.14
Hurd (1968)	*Lycopersicon esculentum* (tomato)	18–43 days (p)	W, L_A	**R, F, E**	
Hughes (1969), Hughes and Cockshull (1969)	*Callistephus chinensis* (China aster)	12 weeks (e)	W, L_A	**R, F, E**	
Goldsworthy (1970)	*Sorghum bicolor* (sorghum)	147 days (p)	W, L_A	**C, L, E**	
Bean (1971)	*Festuca arundinacea* (tall fescue)	28 days (after anthesis)	W	**R**	
Ojehomon (1970)	*Vigna unguiculata* (cowpea)	70 days (p)	Ws, Ns, L_A	**R, F, E, C, L**	
Eckardt *et al.* (1971)	*Helianthus annuus* (sunflower)	43–106 days (p)	W	**C**	
Hughes and Cockshull (1971a, b, c, 1972)	*Chrysanthemum morifolium* (chrysanthemum)	100 days (p)	Ws, L_A	**R, F, E,** SLA, LWR	
Thornley and Hesketh (1972)	*Gossypium* sp. (cotton)	40 days (e)	W	**R**	
Hughes (1973a, b)	*Chrysanthemum morifolium* (chrysanthemum)	100 days (p)	Ws, L_A	**R, F, E,** SLA, LWR	

(*Continued*)

Table 5.3 (*Continued*)

Author(s)	Species	Time interval(s)	Primary data	Derived data	Comments
Hunt and Burnett (1973)	*Lolium perenne* (perennial ryegrass)	58 days (g)	*W*s, L_A, K, *FW*	**Rs, F, E, A**, ratios	
Goldsworthy and Colegrove (1974)	*Zea mays* (maize, 5 vars.)	53–160 days (p)	*W*, **L**	**C, E**	
Goldsworthy, Palmer and Sperling (1974)	*Zea mays* (maize, 3 vars.)	40–40 days (p)	*W*, **L**	**C, E**	
Hurd and Thornley (1974)	*Lycopersicon esulentum* (tomato)	Up to *c.* 5 g total *W*	*W*s, L_A	**Rs, F, E**	**E** 'unrealistic at extreme'
Christie and Moorby (1975)	3 Australian grasses (semi-arid)	46 days (p)	*W*, L_A, P	**R, F, E, A_P**	
Fischer and Wilson (1975)	*Sorghum bicolor* (sorghum)	12 weeks (g)	*W*s, L_A	**R, L, E, C**	
Promnitz (1975)	*Populus* sp. (popular)	12 weeks (g)	*W*	**R**	
Hodgkinson and Quinn (1976)	*Danthonia caespitosa* (tussock grass, 5 vars.)	18–53 days	*W*, *N*, L_A	**Rs, E,** LWR	**E** on basis of L_W
Hunt and Parsons (1977)	*Zea mays* (maize)	113 days (g)	*W*	**R**	See Fig. 5.12
Smith and Neales (1977)	*Prunus* sp. (peach)	113 days (e)	*W*, L_A	**R**	
Mauny, Fry and Guinn (1978)	4 crop spp.	14–119 days (g)	*W*, L_A	**R, E**	
Sorensen (1978)	*Pseudotsuga menziesii* (Douglas fir)	2 years	*H*		
Littleton *et al.* (1979a, b)	*Vigna unguiculata* (cowpea)	90 days (e)	*W*s, **L**	**E**	
Angus *et al.* (1980)	*Triticum aestivum* (wheat)	139 days (e)	*W*s, N	**C, A_N**	

For footnotes see Table 5.1

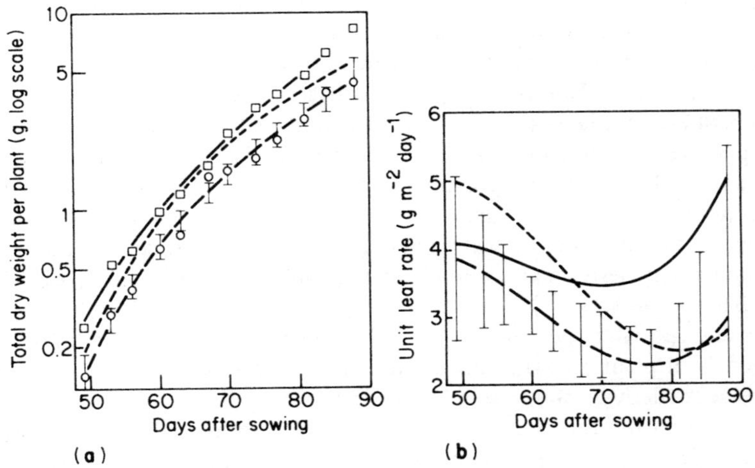

Fig. 5.14 (a) Third-order polynomial exponentials fitted to total dry weight per plant in *Callistephus chinensis* grown in controlled environments at various concentrations of CO_2 : □–□ 900; ––– 600; and ○–○ 325 volumes per million. **(b)** Progressions of unit leaf rate, symbols as in **(a)**, derived ultimately from **(a)** and from similar progressions for total leaf area per plant. 95% limits are shown (data and analyses from Hughes and Freeman, 1967).

5.5 High-order polynomials

5.5.1 The functions and their properties

If we take n, the order of polynomial, as >3 we have

$$W \text{ (or } \log_e W) = a + b_1 T + b_2 T^2 + b_3 T^3 + \ldots + b_n T^n \tag{5.14}$$

from which

$$\frac{dW}{dT} \left(\text{or } \frac{1}{W} \cdot \frac{dW}{dT}\right) = b_1 + 2b_2 T + 3b_3 T^2 + \ldots + n b_n T^{n-1} \tag{5.15}$$

the slope of the nth-order polynomial behaving, naturally, as an $(n - 1)$th-order function. We saw in the case of the third-order polynomial (section 5.4.1) that it was possible only on occasion to ascribe biological significance to its coefficients, so we are not surprised to find here that such a task is invariably impossible. The functions stand purely as empirical equations, useful for modelling progressions of moderate complexity (i.e. beyond the reach of the cubic) and upwards, but quite devoid of any biological basis which could assist with the choice of their form, or, indeed, of their order. Mead (1971) has discussed many of the considerations surrounding the use of this type of function.

Despite the obvious advantages that accrue from their great flexibility, two practical difficulties limit the value of high-order polynomials to plant growth analysis. (1) Computation: during the evaluation of the coefficients of any one function, as the order of each term (the power to which the independent variable is raised) rises, the absolute value of its coefficient normally becomes smaller, and on a geometrical progression. This eventually taxes the numerical powers of even the largest computer (see section 4.6). (2) Over-fitting: a series of n points (or mean values in n normally-distributed arrays of replicated data) may be fitted *exactly* by a polynomial of order $n - 1$. For the reasons given in sections 3.3 and 8.2.4 this is most undesirable on biological grounds. Evidence of both of these problems may be found in the specimen ninth-order polynomial fit performed on some of Kreusler's data (Table 4.1, Fig. 4.6; data from Table 1.1).

In one special respect, however, high-order polynomials are indispensible as models of plant growth. This is where it is necessary to obtain second, or subsequent, derivatives for some comparative purpose. Since one power of T is lost from the function with each successive differentiation, the capacity of low-order polynomials to withstand such a treatment is strictly limited. For example, a cubic fit to $\log_e W$ yields, as we have seen, a quadratic $d(\log_e W)/dT$, and then a linear $d^2(\log_e W)/dT^2$, followed by a zero $d^3(\log_e W)/dT^3$. If flexibility is to be retained in these latter derivatives, more powers of T are called for at the outset. One striking example of this, though outside plant growth analysis, is Joossens and Brems-Heyns's (1975) analysis of the human growth curve – they produced near-ultimate flexibility in derivates representing velocity and acceleration by employing polynomials of up to the *eighteenth* order in fits to their primary data. This was done to obtain comparative representations, for different populations, of such phenomena as the 'adolescent growth spurt'.

5.5.2 Case studies

Work involving high-order polynomial fits to plant data is listed in Table 5.4, which differs from previous tables in this chapter only in that the order(s) of polynomial employed is(are) identified alongside the primary data.

The publications in which derived data have been obtained from high-order polynomials can each be recommended as an example of the methodology and, of these, Hall (1977) provides perhaps the simplest example of how great flexibility may be retained in first derivatives, in this case quartic progressions of **R** (Fig. 5.15).

5.6 Stepwise polynomials

5.6.1 The stepwise principle and its consequences

A stepwise polynomial is one which has been constructed by the successive addition (or, occasionally, subtraction) of polynomial terms until the point where, as a result of continuous re-assessment of the whole function, the

Table 5.4 Applications of high-order polynomials (untransformed primary data) and high-order polynomial exponentials (logarithmically transformed primary data).

Author(s)	Species	Time interval(s)	Primary data (and order(s) of polynomial)	Derived data	Comments
Untransformed primary data					
Koller, Nyquist and Chorush (1970)	*Glycine max* (soybean)	100 days (p)	*W*s, **L** (5, 6, 7)	**C, R, F, E** SLA, LWR	
Grace and Woolhouse (1973)	*Calluna vulgaris* (heather)	6 months (e)	L_W (4, 5)	**G**	
Koller (1971)	*Glycine max* (soybean)	90 days (p)	*W*s, L_A (6, 7)	**C, R**s	
Hall (1977)	*Capsicum annuum* (green pepper)	20 days before to 100 days after anthesis	*W*s, L_A (up to 5)	**R**s, **F, G, E**	
Bayly and Shibley (1976)	*Pontederia cordata* (a N.Amer, fresh water macrophyte)	82 days (e)	Ca, K, Mg, Na contents (3, 4)		Many computer produced Figs.
Christy and Fisher (1978)	*Ipomea nil* (morning glory)	8 hours (e)	Translocation rate of ^{14}C (8)		
Griggs, Nance and Dinus (1978)	*Pinus elliottii* (slash pine)	12 years (e)	Percent infected individuals (4)		Disease progress curves
Acevedo *et al.* (1979)	*Sorghum bicolor* (sorghum), *Zea mays* (maize)	16 hours (diurnal trends)	Leaf solute potential (4)		
McKee *et al.* (1979)	*Avena sativa* (spring oats)	40 days (after anthesis)	Grain *W* (5)		
Logarithmically transformed primary data					
Goodhall (1949)	*Theobroma cacao* (cacao)	47 days (p)	*W*s, L_A H_2O content (up to 5)		Many valuable Figs.
Hackett and Rawson (1974)	*Nicotiana tabacum* (tobacco)	26–90 days (p)	*W*s, L_A (up to 4)	**R**s, SLA, **E**	

For footnotes see Table 5.1

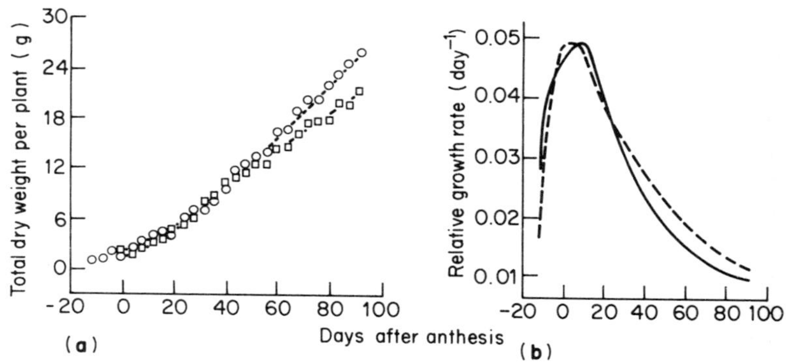

Fig. 5.15 (**a**) Fifth-order polynomial fits to total dry weight per plant in fruiting (□—□) and deflorated (○– –○) plants of *Capsicum annuum*; (**b**) relative growth rates derived from (**a**) (data and analyses from Hall, 1977).

computer program, or experimenter, decides to call a halt. Polynomial functions settled upon by this method differ neither in themselves nor in their properties from the functions listed in the preceding sections of this chapter: it is the route by which they are selected that is special.

This route may be taken in one of three ways: (1) by fitting single polynomials manually, and independently, but in a series sufficiently lengthy to encompass the principal contenders among orders of polynomial; (2) by employing a commercially available method of stepwise polynomial regression, such as the program BMD02R of Dixon (1973); and (3) by employing a stepwise program purpose-made for plant growth analysis, such as that of Nicholls and Calder (1973) or of Hunt and Parsons (1974), both of which see.

Unfortunately, or perhaps fortunately for experimenters who wish to retain the maximum degree of control in their own hands, there is no universally accepted criterion for settling upon a stopping point when performing a stepwise regression. Any of three different methods may be of value. (1) The deviations of the primary data about the fitted regression may be minimized, leaving outside the regression only that portion of the variance believed to be random in origin. This normally results in minimal standard errors for fitted values of the dependent variate and is a particularly useful method for selecting suitable *multiple* regression models (that is, models involving more than one independent variate – section 7.7 – and in which variables or terms, once admitted, may subsequently be removed). But, in the context of plant growth analysis, where, of course, time alone is our abscissa and terms, once admitted, are permanent, the method may well lead to 'over-fitting', that is, to too little smoothing. (2) The statistical significance of the most lately arrived polynomial coefficient, b_n, may be examined individually and the series halted at b_{n-1} should b_n prove not to be significant at, say $P<0.05$. This method is useful in the lower part of the polynomial series where experimenters value the coefficients themselves for their biological information. For example, a b_3 coefficient may be necessary to find out whether or not there is any suggestion of an acceleration in, say, a decline in **R**; for, in borderline cases, this coefficient may be statistically significant when some other criterion suggests that none higher than b_2 should be admitted. (3) Then, the overall significance of the regression may be assessed by way of Fisher's test, which computes an F-value (here, the ratio between the deviations of means mean square and the residual mean square). This statistic takes into consideration the proportion of the total variance in the primary data which is, and is not, accounted for by the regression equation, and also allows for the number of parameters currently in the equation and the number of original data. In this way it is possible to avoid a further step beyond some present level, n, even if leading to a significant b_{n+1} and a reduced residual, if the loss of one degree of freedom (through the addition of another parameter) would lead to diminished returns in the size of the F-value. Compared with the other two methods, this approach tends towards conservatism in allocating coefficients when data are few. Finally, it should be said that there will be many instances in which these three criteria will each yield precisely the same result.

The stepwise principle is, quite wisely, used to protect data from the abuses that result either from under-fitting or from over-fitting. Of the former, we already have a perfectly satisfactory example in the log-linear fit to Kreusler's data shown in Fig. 5.2 while, of the latter, the two pairs of progressions of **E** included by Hunt and Parsons (1974) as part of their attack on the over-fitting frequently encountered in the uncritical use of Hughes and Freeman's (1967) method, illustrate that needless trouble may easily be avoided (Fig. 5.16).

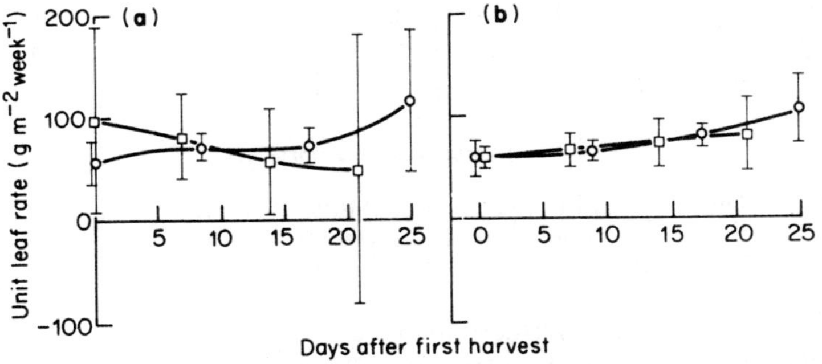

Fig. 5.16 Progressions of unit leaf rate in seedlings of *Holcus lanatus* (○) and *Nardus stricta* (□); **(a)** derived ultimately from third-order polynomial exponentials fitted to the primary data; **(b)** derived ultimately from first- (*Nardus*) or second-order (*Holcus*) polynomial exponentials fitted to the same primary data (data and analyses from Hunt and Parsons, 1974).

But one disadvantage of stepwise polynomials, especially when operating low in the range of orders, is that difficulties sometimes occur when trying to interpret a whole body of data to which regressions have been applied objectively according to the needs of the individual sets. Because the form of the progressions of the derived quantities on time depends immutably upon the nature of the original regressions, and because small differences among the latter are often reflected in large differences among the former, comparisons between species or treatments are sometimes puzzling when quite distinct patterns in the behaviour of, say, **R** or **E** emerge from sets of raw data which differ only slightly from one another, but nevertheless enough to cause the selection of different growth functions.

This problem was highlighted by Hurd (1977) who grew *Lycopersicon esculentum* (tomato) under solution-culture conditions in a controlled environment and in each of eight treatments fitted the logarithms of W and L_A by Hunt and Parsons's (1974) method. Of the sixteen resulting regressions, four were linear, six were quadratic and six were cubic (Fisher's test, $P<0.05$). Hurd argued that this selection was unsatisfactory on biological grounds and that quadratic models should be used throughout to give a uniform family of curves.

One way of deliberately avoiding such a mixed bag of growth functions is to stiffen the requirements for their acceptance. The relationship between the probability level for acceptance of the models and the types of model chosen was calculated for Hurd's sixteen sets of data by Hunt (1978a) (Fig. 5.17).

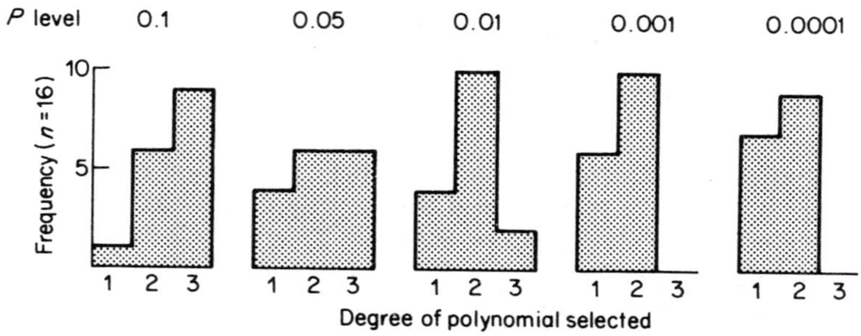

Fig. 5.17 The effect of varying the probability levels required for the acceptance of polynomial exponential fits in sixteen sets of data on the growth of tomato (data from Hurd, 1977; analyses from Hunt, 1978a).

The result of raising the statistical threshold for the choice of polynomial was a forcing-out of the higher-order terms. That is, as the requirements for acceptance become progressively more stringent, cubics turned into quadratics and quadratics turned into straight lines. But at no time was a single model unquestionably the best. Hurd's suggestion that the quadratic regression might be used throughout therefore needs to be treated with caution – as a semi-mechanistic model it has its advantages but these accrue at the cost of accuracy of fit.

So, when fitting stepwise polynomials the experimenter can well be faced with two requirements which conflict: statistical exactitude, and what Hurd (1977) called the experimenter's 'biological expectation'. A rigid adherence to either is unwise: to statistical exactitude because the resulting progressions of fitted primary or derived data may be biologically unrealistic, or even impossible, and to biological expectation because semi-mechanistic requirements are often quite unsupported by the data themselves, as when what is believed to be the reality of growth is submerged by careless or unlucky sampling or by the natural variability or inaccessibility of the subject material. It is not a question of the degree of sophistication of the model, for statistical exactitude may suggest a model either more or less complicated than biological expectation. It is a question of balance. If the experimenter can justify either as being of overriding importance then the other can be ignored. But in the more usual case where both statistical exactitude and biological expectation are given some credence, the experimenter himself can be the only arbiter. Should there be a dilemma, one flash of Occam's razor will remove its cause (Nicholls and Calder, 1973, Hunt and Parsons, 1974).

5.6.2 Case studies

Work involving stepwise polynomials is listed in Table 5.5. The criterion for inclusion here has been the authors' consideration of at least three orders of polynomial simultaneously (and by one of the three methods outlined in section 5.6.1), whether or not only one order was ultimately selected for use throughout.

Naturally, authors employing the logarithmic transformation have been less reticent than others about proceeding to derivates from their curves. This is partly because much of the more recent work has involved the computer program developed by Hunt and Parsons (1974). This incorporates a stepwise polynomial exponential fit, from the first to the third order, which is adjusted to accept the simplest possible models independently for two primary variates. Fitted values of these variates are then calculated, with their 95% limits, and then appropriate derivates are obtained in types (i), (ii) and (iii) (see section 2.1), again, all with 95% limits.

Many workers employing stepwise polynomials have advanced their own opinions on the value of the method and of the results obtained from it. So, in addition to referring to the works already cited in this section, readers should note the points variously made by Buttery and Buzzell (1974), Grime and Hunt (1975), Elias and Causton (1976), Neales and Nicholls (1978) and Clarke and Simpson (1979).

Table 5.5 Applications of stepwise polynomials (untransformed primary data) and stepwise polynomial exponentials (logarithmically-transformed primary data).

Author(s)	Species	Time interval(s)	Primary data	Derived data	Comments*
Untransformed primary data					
Christiansen (1962)	*Gossypium hirsutum* (cotton)	5 days (g)	Percent cotyledon *W*		Orthogonal polynomials
Elliott and Jardine (1972)	*Triticum aestivum* (wheat)	29 years	*W* ha^{-1}		Long-term trends in yields
Bayly and Shibley (1978)	*Pontederia cordata* (a N. Amer. fresh water macrophyte)	82 days (e)	Ca, K, Mg, Na contents		Many computer-produced figs
Boot *et al.* (1978)	*Glycine max* (soybean)	35 days (e)	N, P, K %, CO_2 influx		
Griggs, Nance and Dinus (1978)	*Pinus elliottii* (slash pine)	12 years (e)	Percent infected trees		Disease progress curves
McCollum (1978)	*Solanum tuberosum* (potato)	84 days (g)	*Ws*, L_A	**Gs, D, E**	
Robson and Parsons (1978)	*Lolium perenne* (perennial ryegrass)	25–125 days	*W*, L_A	**Rs, E, F**	
Kalmbacher, Minnick and Martin (1979)	*Trifolium* spp. (clover), *Medicago sativa* (alfalfa)	29 days (e)	Percent mortality		Grazing by snails
Patterson, Raper and Gross (1979)	*Glycine max* (soybean)	18 days (e)	C_2H_4 reduction rates, *Ws*, pod *N*		
Semu and Hume (1979)	*Glycine max* (soybean)	120 days (p)	N_2 fixation, *Ws*		
Young, Whisler and Hodges (1979)	*Glycine max* (soybean)	108 days (p)	Leaf N		
Karlen and Whitney (1980)	*Triticum aestivum* (winter wheat)	110 days (p, or after dormancy)	*Ws*, 9 mineral elements		Many computer-produced figs

Table 5.5 (*Continued*)

Author(s)	Species	Time interval(s)	Primary data	Derived data	Comments*
Willms, McLean and Kalnin (1980)	6 British Columbia grasses	106 days (e)	Fibre, lignin and protein percentages		
Pulli (1980c)	Grassland community	Spring–Autumn	W	**C**	
Logarithmically transformed primary data					
Mather (1964)	*Lycopersicon esculentum* (tomato)	45 days (g)	W		Data from Ashby (1937), orthogonal polynomials
Silsbury (1969)	*Lolium* spp. (ryegrass)	32 days (p)	Ws	**R**	
Davies (1971)	*Lolium perenne* (perennial ryegrass)	28 days (e)	S_W	**R**	
Nicholls and Calder (1973)	*Atriplex* spp. (orache)	40 days (p)	W, L_A	**R**s, LWR, **E**	
Buttery and Buzzell (1974)	*Glycine max* (soybean)	84 days (p)	W, L_A	**E**	
Callaghan (1974)	*Phleum alpinum* (alpine cat's-tail)	69 days (e)	W, L_A	**R**s, **F**, **E**	
Hunt and Parsons (1974)	*Holcus lanatus* (Yorkshire fog) *Nardus stricta* (mat-grass)	12–35 days (g)	W, L_A	**R**s, **F**, **E**	See Figs 5.16, 5.17; data of Grime and Hunt (1975)
Rorison and Gupta (1974)	*Rumex acetosa* (sorrel), *Scabiosa columbaria* (small scabious)	28 days (g)	Ws	Ratios	H&P (1974)
Grime and Hunt (1975)	132 spp., mainly British natives	14–35 days (g)	W	**R**	H&P (1974)
Hunt (1975)	*Lolium perenne* (perennial ryegrass)	37 days (g)	Ws, nutrient contents	**R**s, ratios, **A**, **B**	H&P (1974)

(*Continued*)

Table 5.5 *(Continued)*

Author(s)	Species	Time interval(s)	Primary data	Derived data	Comments*
Hunt, Stribley and Read (1975)	*Vaccinium macrocarpon* (cranberry)	20 weeks (g)	*W*s, N contents	Ratios, **A, B**	H&P (1974)
Stribley and Read (1975)	*Vaccinium macrocarpon* (cranberry)	15 weeks (p)	*W*s, N contents		H&P (1974)
Stribley, Read and Hunt (1975)	*Vaccinium macrocarpon* (cranberry)	20 weeks (p)	*W*s, N contents	**R**, %N, *R/S*, **A**	H&P (1974)
Callaghan (1976)	*Carex bigelowii* (stiff edge)	7 years	Tiller *W*	– **R**	
Elias and Causton (1976)	*Blackstonia perfoliata* (yellow-wort), *Daucus carota* (carrot), *Impatiens parviflora* (small-leaved balsam)	14 weeks (g)	W, L_A		
Stribley and Read (1976)	*Vaccinium macrocarpon* (cranberry)	21–84 days (p)	*W*s, N contents	**R**s, ratios, **A**	H&P (1974)
Walton and Smith (1976)	*Avena sativa* (oat)	110 days (e)	*N*s, L_A, *W*s	**R, F** and **E** determined classically	
Bazzaz and Harper (1977)	*Linum usitatissimum* (flax)	108 days (g)	*W*s, L_A	**R**s, **F, E**	H&P (1974)
Hunt and Parsons (1977)	*Zea mays* (maize)	113 days (g)	*W*	**R**	Data from Table 1.1, see Fig. 5.12, H&P (1974)
Hurd (1977)	*Lycopersicon esculentum* (tomato)	40 days (p)	W, L_A	**R**s, **F, E**	H&P (1974)
Jarvis and Wilson (1977)	*Corylus avellana* (hazel)	18 days (g)	Embryonic axis *L*		H&P (1974)
Morgan and Parbery (1977)	*Medicago sativa* (lucerne)	21–53 days (p)	W, L_A	**R**s, **F, E**	

Table 5.5 *(Continued)*

Author(s)	Species	Time interval(s)	Primary data	Derived data	Comments*
Sanders *et al.* (1977)	*Allium cepa* (onion)	50 days (p)	% infection, R_L, R_P	**A**	H&P (1974)
Crittenden and Read (1978)	*Lolium perenne* (perennial ryegrass)	16 weeks (p)	Leaf, tiller N, *W*s	**R**s	H&P (1974)
Jarvis and Wilson (1978)	*Corylus avellana* (hazel)	14 days (g)	Embryonic axis *L*		H&P (1974)
Jarvis, Wilson and Fowler (1978)	*Corylus avallana* (hazel)	18 days (g)	Embryonic axis *L*		H&P (1974)
McCollum (1978)	*Solanum tuberosum* (potato)	84 days (g)	W, L_A	**R**s, **D**, **E**	
Neales and Nicholls (1978)	*Triticum aestivum* (wheat)	24 days (g)	Ws, L_A	**R**s, **F**, **E**	H&P (1974)
Abul-Fatih, Bazzaz and Hunt (1979)	*Ambrosia trifida* (giant ragweed)	36–97 days (g)	W, L_A	**R**s, **F**, **E**	H&P (1974)
Brewster (1979)	*Allium* spp. (onion, leek, chives)	About 100-fold increase in *W*	Ws, L_A	**R**s, **F**, **E**,	
Clarke and Simpson (1979)	*Brassica napus* (rape)	60 days (e)	W, L_A	**C**, **L**	
Crittenden and Read (1979)	*Lolium multiflorum* (Italian ryegrass), *Dactylis glomerata* (cock's foot)	70 days (p)	S_W	$\mathbf{R_S}$	
Elias and Chadwick (1979)	40 commercial grasses and legumes	14–35 days (g)	*W*s	**R**s, ratios	H&P (1974)
Jarvis (1979)	*Corylus avellana* (hazel)	14 days (g)	Embryonic axis *L*		H&P (1974)
Angus *et al.* (1980)	*Triticum aestivum* (wheat)	110 days (p)	*W*s, **L**, N, P	**A**, **C**	
Horsman, Nicholls and Calder (1980)	3 grasses	36 days (g)	Ws, L_A	**R**, **F**, **E**	

(Continued)

Table 5.5 (*Continued*)

Author(s)	Species	Time interval(s)	Primary data	Derived data	Comments*
Hunt and Bazzaz (1980)	*Ambrosia trifida* (giant ragweed)	36 days (g)	Ws, L_A	**R**s, **F**, **J**	H&P (1974)
Middleton, Jarvis and Booth (1980)	*Phaseolus aureus* (mung bean)	4 days (e)	R_W, R_L, R_N		H&P (1974)

*'H&P (1974)' indicates that the computer program described by Hunt and Parsons (1974) was employed.
For other notes see Table 5.1.

6

Asymptotic functions

6.1 Introduction

In an asymptotic function, whatever its other properties, the form of its progression is governed by one characteristic feature: the value of the dependent variate more and more gradually ascends (or descends) to a plateau which it never quite meets. This plateau in the value of Y is known as the asymptote, or asymptotic value, and is the value predicted for Y when X is at infinity. The name is derived from the Greek ἀσύμπτωτος, meaning 'not falling together' (ἀ-, not; σύμ (for σύν), together; and πτωτός, falling).

Though this property of asymptotic functions is universal, it is not necessarily obvious in the form of every fitted progression. For example, when data, though fitted by an asymptotic function, do not include a marked approach to a plateau within their range, then the resulting progression fitted to this range would not be likely, on inspection alone, to appear asymptotic. However, in such a case the asymptotic value present in the equation of the fitted curve will always become graphically evident if sufficient extrapolation is performed. Then, in addition to this terminal feature, whether obvious or not, asymptotic functions may also vary widely in the forms of the early and middle parts of their progressions, according to which one of this wide family of functions is being employed.

Unlike the polynomial functions described in Chapter 5, these asymptotic functions are statistically non-linear. Put simply, this means that their parameters, which commonly number three or more, are not additively combined in a linear sequence (of whatever length) as here:

$$\log_e W = a + b_1 T + b_2 T^2 + b_3 T^3 + \ldots + b_n T^n, \tag{5.5}$$

but are subject to division, multiplication or exponentiation with one other, as in

$$W = a(1 - b\mathrm{e}^{-cT}), \tag{3.3}$$

the monomolecular function, first encountered in section 3.2 and to be discussed further in section 6.2. This type of structure means that no simple and direct methods exist for obtaining estimated values of parameters from a set of observational data. Instead, the computational procedure normally followed involves the assignment of arbitrary starting values (that is, values derived by judgement not calculation) to one or more of these parameters (guided or not by what is already known about the data in advance), the solution of this provisional equation, and the subsequent adjustment of the parameter value(s) in the hope of achieving a closer fit to the data on a second attempt. By this method, the parameter values gradually converge upon those of the equation which gives the best possible fit to the data in hand, and do so by way of what may well be a lengthy series of iterations (for which full-scale computing facilities, and a program such as the BMD06R of Dixon (1973) are a great advantage).

Though various forms of asymptotic function have been in use for a considerable period of time, such computational difficulties have limited their use largely to their being only representational models of plant growth, without the full statistical apparatus attached to the fitted progressions and their derivates which is so vital to plant growth analysis. For example, F. J. Richard's function (section 6.5), first described by him in 1959, had to wait for twenty years (even in the 'computer age') until this necessary supplementation could be provided (by Venus and Causton, 1979c).

Many of the asymptotic functions found their way into plant science through their earlier successes in modelling the growth of animals. We have already seen (section 3.2) how one vital difference between animals and plants – that of animals coming into the world with both the numbers and (within limits) the sizes of their organs genetically pre-determined, and that of plants not so, or hardly at all – influences and constrains the experimenter seeking to fit growth curves. Because the growth of whole plants is much less determinate than that of whole animals, various combinations of genetical and environmental influences can produce a wide range of natural end-points to the growth curve. In whole plants, the very concept of a final, limiting size is suspect, and to predict it from juvenile performance, as is necessary when dealing with data which span less than the whole life history of the plant, can be misleading. Final size in whole plants is a result, not a goal; in animals it is both.

However, when dealing with plant organs, the situation is quite different. Often we see that once a certain size has been attained as a result of the development of a single root, leaf or flower primordium, further growth of the whole organism (if any is possible) occurs not as increases within the same organ, but as the adaition of, and increases within, subsequent organs (see, for example, Fig. 6.13). Such processes led Bazzaz and Harper (1977) to suggest that since

> 'By far the greater part of plant growth is due to the addition of modules that are of *comparatively* constant size . . . the plant is a population of parts which is perfectly adapted for [*sic*] the study of age structure'.

Such studies involve the demographic derivates of birth and death rates, survivorship and life expectancy. Hunt (1978b) argued against this notion and Hunt and Bazzaz (1980) showed that the concept of the 'repeatable module' was by no means invariable and that plants could be plastic both in the number of their organs and in their ultimate size. This argument now lies in a position of stalemate, convincing examples of both interpretations having been found. But the part of it which is relevant here is that, whether whole plants are pre-determined in size or not, asymptotic functions provide encouraging possibilities for the modelling of trends in organs, the sizes of which ultimately do approach plateaux. Great play is made of this property by Causton and Venus (1981), who conclude their work with a whole plant model synthesized (with statistical limits) from empirical asymptotic fits to data on the growth of the plant's individual organs.

Polynomial functions have no satisfactory way of dealing with asymptoticism. If splined (section 7.4), or if of sufficiently high order (section 5.5), they may make a reasonable attempt at approaches to plateaux, but this counts as clumsiness against the performance of asymptotic functions especially constructed for this purpose. If the progression of primary data though, reaches a plateau ***and then falls again*** (as in Fig. 3a of Parsons and Hunt, 1981), this position is dramatically reversed and it is then the turn of the asymptotic function to become the fish out of water, decline from a maximum not being a possibility for any function of this type.

This chapter will deal one-by-one with asymptotic functions which have most commonly been used in plant growth analysis (and will conclude with the briefest mention of others). This will be done at a depth of coverage sufficient only to achieve the essentials of uniformity with Chapters 5 and 7, since further information on this special topic is readily available from the aforementioned and related volume by Causton and Venus (1981). In addition, comparative information on various asymptotic functions has been assembled by Richards (1969), Causton (1970), Analytis (1974), Lioret (1974), Hasegawa (1976), Grimm (1977), Landsberg (1977), Obrucheva and Kovalev (1979), Pruitt, DeMuth and Turner (1979), Milthorpe and Moorby (1980), and Żelawski and Lech (1980).

6.2 Monomolecular function

6.2.1 The function and its properties

The monomolecular function, borrowed into biology from physical chemistry (section 3.2), is, as we have seen,

$$W(\text{or } \log_e W) = a(1 - b\mathrm{e}^{-cT}) \qquad (3.3)$$

where a is the asymptote, b a measure of the starting size of the system and c is a rate constant. Equation 3.3 is the basic function fitted to untransformed primary data. In asymptotic functions, as in the case of polynomials, the question of logarithmic transformation often arises. However, unlike polynomials, asymptotic functions can also be subjected to logarithmic transformation for the mathematical purpose of linearizing or simplifying the function to secure an easier method of fitting. This can be in addition to, or instead of, the separate statistical objective of ensuring homoscedasticity of variance. Transforming *both* sides of equation 3.3 to logarithms produces

$$\log_e W = \log_e a + \log_e (1 - b\mathrm{e}^{-cT}) \,. \tag{6.1a}$$

Here, the parameters retain their original relationships *vis-à-vis* with the un-transformed data and the growth function itself becomes simpler, with changed properties (see Causton, 1977, pp. 197–202).

Transforming the left-hand side *only* of equation 3.3 gives

$$\log_e W = a(1 - b\mathrm{e}^{-cT}) \tag{6.1b}$$

an equation which, following the terminology adopted for polynomials by Causton (1970), may be termed the monomolecular exponential. This may be done in cases where the *logarithms of* the primary data are homoscedastic and lie in a progression, the general shape of which corresponds broadly to that of the function. The structure and properties of the function are not altered by this transformation: it is merely applied to the logarithms of data instead of to their arithmetic values, back-transformation of fitted data being necessary if the experimenter wishes for any reason to perform subsequent comparisons on an un-transformed basis. Values of $\log_e W$ may require the addition of a constant to remove negativity, and this must also be allowed for in a.

There is an important difference between this procedure and that in which the *whole* function is transformed to logarithms as in equation 6.1a. In this former case, if the experimenter is prepared to estimate and iterate around the asymptotic value, a, the basic monomolecular function written in the form of equation 6.1a may be linearized and fitted to data in the form

$$\log_e [1 - (W/a)] = \log_e b - cT \tag{6.2a}$$

and the same may be done for the monomolecular exponential, equation 6.1b:

$$\log_e [1 - (\log_e W/a)] = \log_e b - cT \,. \tag{6.2b}$$

Linearization by logarithmic transformation is, in equations 6.2a and 6.2b, a mathematical device which brings the non-linear function within the bounds of simple regression methodology. But it also alters the statistical properties of the primary data and the experimenter should be aware of whether or not these alterations have been harmful. For example, plotted against time, untransformed data may well show increasing variability, transformed data may show uniform variability (the ideal situation), and doubly transformed data may show decreasing variability. Be aware also that the identity of the desired derivate (absolute or relative growth rate) need not affect the fitting procedure, since both are available from either method.

Following the scheme outlined in section 3.4, these derivates are, for the monomolecular (equation 3.3):

$$\frac{\mathrm{d}W}{\mathrm{d}T} = abce^{-cT} \tag{6.3a}$$

and

$$\frac{1}{W} \cdot \frac{\mathrm{d}W}{\mathrm{d}T} = \frac{abce^{-cT}}{a(1 - be^{-cT})} = \frac{bce^{-cT}}{1 - be^{-cT}} \tag{6.3b}$$

and for the monomolecular exponential (equation 6.1b):

$$\frac{\mathrm{d}W}{\mathrm{d}T} = abce^{-cT} \cdot e^{[a(1 - be^{-cT})]} = abce^{[a(1 - be^{-cT}) - cT]} \tag{6.3c}$$

and

$$\frac{1}{W} \cdot \frac{\mathrm{d}W}{\mathrm{d}T} = abce^{-cT} \ . \tag{6.3d}$$

The monomolecular is one of the simplest of asymptotic functions. It has no point of inflection and its slope has a progression which is convex to the time axis, being proportional to the amount of growth yet to be made, $a - W$ (Fig. 6.1). In this respect it can be useful in the same situations as one of the variants of the third-order polynomial (Fig. 5.11c), though it differs from that function in that, being genuinely asymptotic, it is unable to proceed into negative slope. Allen (1976) discusses its properties further.

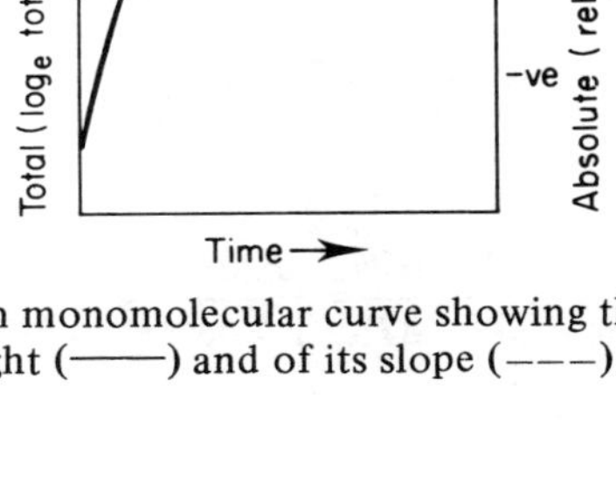

Fig. 6.1 A specimen monomolecular curve showing the progressions of total or $\log_e$ total dry weight (——) and of its slope (– – –).

6.2.2 Case studies

The lack of inflection has almost certainly limited the usefulness of this function, even for the modelling of the determinate growth of individual organs, though one early success of this type was gained by Gregory (1928b). However, in cases which a progression is either constrained to be (or ultimately turns out to be) asymptotic *and*, by dint of starting late in the life of the system, it has no inflection, then the monomolecular function may be appropriate. Such a combination of events was encountered by Blacklow and McGuire (1971) who investigated *re*-growth, following clipping, from a sizeable existing biomass of *Festuca arundinacea* (tall fescue) growing in a relatively closed canopy. They presented convincingly linear plots of their data in the form of equation 6.2a. Homès and van Schoor (1978) segmented monomolecular fits with first-order polynomial exponentials (section 7.2) over periods of 180 days when studying nutrient contents of *Beta vulgaris* (sugar beet).

6.3 Logistic function

6.3.1 The function and its properties

This function is, like the monomolecular, a three-parameter function. It is also known as the autocatalytic function and takes the form

$$W \text{ (or } \log_e W) = a/(1 + b\mathrm{e}^{-cT}) \tag{6.4}$$

where the bracketed term on the left-hand side refers to the dependent variate of the logistic exponential. The value of W (or $\log_e W$) at $T = 0$ is $a/(1 + b)$ and the function has a symmetrically placed point of inflection at $T = (\log_e b)/c$ and W (or $\log_e W$) $= a/2$. After linearization the functions may be fitted in the form

$$\log_e [(a/W) - 1] = \log_e b - cT \tag{6.5a}$$

or

$$\log_e [(a/\log_e W)] - 1 = \log_e b - cT \ . \tag{6.5b}$$

For the logistic, the slopes are

$$\frac{\mathrm{d}W}{\mathrm{d}T} = \frac{abc\mathrm{e}^{-cT}}{(1 + b\mathrm{e}^{-cT})^2} \tag{6.6a}$$

and

$$\frac{1}{W} \cdot \frac{\mathrm{d}W}{\mathrm{d}T} = \frac{bc\mathrm{e}^{-cT}}{1 + b\mathrm{e}^{-cT}} \tag{6.6b}$$

and for the logistic exponential

$$\frac{dW}{dT} = \frac{abc\mathrm{e}^{[a/(1+b\mathrm{e}^{-cT}) - cT]}}{(1 + b\mathrm{e}^{-cT})^2} \tag{6.6c}$$

and

$$\frac{1}{W} \cdot \frac{dW}{dT} = \frac{abc\mathrm{e}^{-cT}}{(1 + b\mathrm{e}^{-cT})^2} \tag{6.6d}$$

A specimen of the function and its slope is shown in Fig. 6.2.

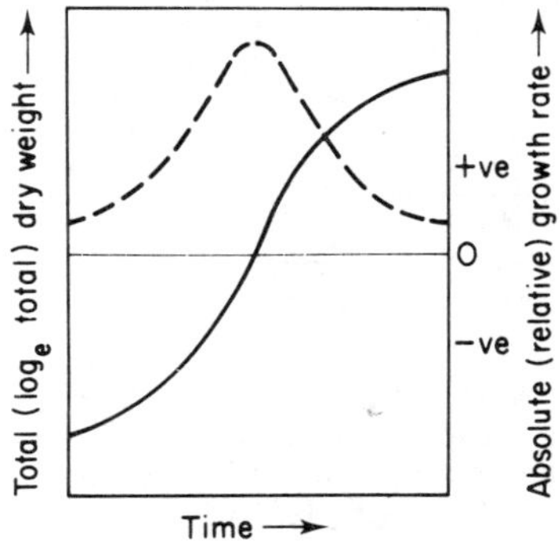

Fig. 6.2 A specimen logistic curve showing the progressions of total or $\log_e$ total dry weight (——) and of its slope (– – –).

6.3.2 Case studies

The logistic equation, of course, has been used very extensively in the field of animal ecology for the modelling of change in numbers of individuals within a population (see Solomon, 1976, p. 14 ff. for an introduction). Biological significance has been attached to parameter c, the 'intrinsic rate of increase' of the population, and to the asymptote, a, the 'carrying capacity of the environment'. More recently, these two parameters have enjoyed a further fame (or notoriety, according to viewpoint) as the basis of 'r-K selection' (MacArthur and Wilson, 1967; Pianka, 1970), r being the usual designation of parameter c in this field, and K that of a.

In plant growth studies, however, the fact that the function is S-shaped (albeit inflecting rigidly at $a/2$) has rendered it very popular. It is a pleasure to be able to illustrate its use with two examples which data back almost to the beginnings of plant growth analysis: Gregory's (1921) fit to the growth of a single leaf of cucumber (Fig. 4.5) and Prescott's (1922) analysis of flower numbers in cotton (Fig. 6.3a) which proceeded as far as absolute growth rate in flower number (Fig. 6.3b). Though these figures lack their primary data, Prescott's separate evidence, given in the form of simultaneous progression of **G** and harvest-interval values of $\overline{\mathbf{G}}$, shows that the curves fitted his data very closely (see Fig. 8.2).

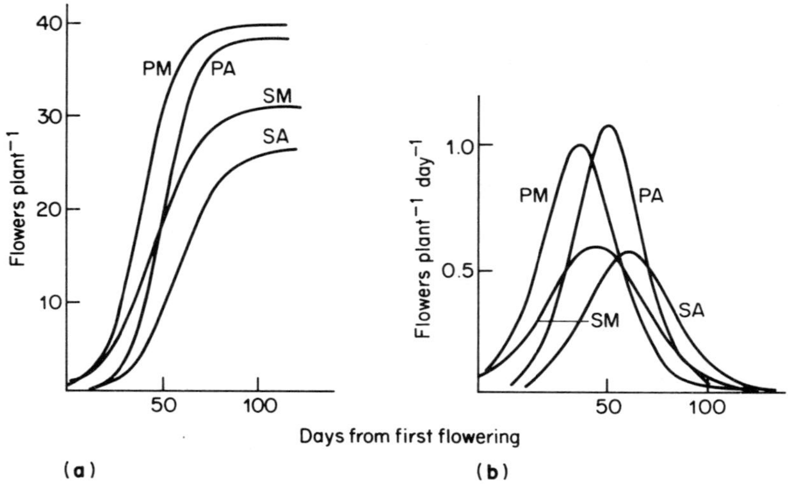

Fig. 6.3 Logistic flowering curves for cotton grown at Bahtim (Egypt) in 1920; P, variety Pilion; S, variety Sakellaridis; M, sown in March; A, sown in April: **(a)** shows flower numbers per plant; the derivate, **(b)**, shows absolute growth rates in flower numbers per plant (data and analyses from Prescott, 1922). See also Fig. 8.2.

One advantage of relatively simple asymptotic functions is that the individual members of families of related curves may each be fitted by the function without losing their family identity. This property is only occasionally shared by collections of polynomials of comparable complexity, such as the third-order, and almost never by stepwise polynomial packages (section 5.6). An example of this happy state of affairs is provided by Silsbury and Fukai's (1977) analysis (Fig. 6.4) of the accumulation of shoot dry matter in *Trifolium subterraneum* (subterranean clover) where, despite involving three times of sowing and three densities of planting, the generic connection between the progressions is never submerged. Also illustrated here is the question of the 'distant asymptote', mentioned in section 6.1. Further applications are given in the extensive list which appears in Table 6.1.

6.4 Gompertz function

6.4.1 The function and its properties

This function, devised by Benjamin Gompertz in 1825, also has three parameters, but these are arranged as a double exponent:

$$W \text{ (or } \log_e W) = a\mathrm{e}^{-b\mathrm{e}^{-cT}}. \tag{6.7}$$

The value of W (or $\log_e W$) at $T = 0$ is $a\mathrm{e}^{-b}$. Like the logistic, the Gompertz's

Table 6.1 Applications of the logistic function.

Author(s)	Species	Time interval	Primary data	Derived data	Comments
Gregory (1921)	*Cucumis sativus* (cucumber)	16 days (p)	L_L, L_A	G	See Fig. 4.5
Prescott (1922)	*Gossypium hirsutum* (cotton)	110 days (from flowering)	Flower N	G	See Figs 6.3, 8.2
Vyvyan (1924)	*Phaseolus vulgaris* (kidney bean)	15 days (g)	L_A		
Gaines and Nevens (1925)	*Helianthus annuus* (sunflower)	80 days (p)	L_A, stem fibre, Al, Ws		
	Zea mays (maize)				
Rudolfs (1927)	*Asparagus officinalis* (asparagus)	65 days (e)	Stem L		
Copeman (1928)	*Vitis vinifera* (grape)	70 days (e)	Sugar content		
Pearl, Winsor and Miner (1928)	*Cucumis melo* (cantaloup melon)	16 days (g)	Stem L		
Porterfield (1928)	*Phyllostachys* spp. (bamboo)	40 days (e)	Shoot L		
Bond (1945)	*Camellia thea* (tea)	25 weeks (p)	Leaf N, L		Includes plastochrons, segments, see sections 2.4, 7.2
Ashby and Wangermann (1950)	*Ipomea purpurea* (morning glory)	16 days (e)	L_A	G	Individual leaves
Brougham (1956)	N.Z. pasture spp.	18 weeks	W	G	
Friend (1960)	*Triticum aestivum* (wheat)	28 days (e)	Leaf chlorophyll	G	
Williams (1964)	*Triticum aestivum* (wheat)	25 weeks (p)	S_W	R	Compared with cubic and segmented fits
Sváb, Mándy and Bócsa (1968)	*Cannabis sativa* (hemp)	13 weeks (p)	H	G	

(Continued)

Table 6.1 (*Continued*)

Author(s)	Species	Time interval	Primary data	Derived data	Comments
Morley (1968)	N.Z. pasture spp.	80 days (p)	W	**G**	
Ledig (1969), Ledig and Perry (1969)	*Pinus taeda* (loblolly pine)	24 weeks (e)	W, L_W	**E**	
Hunt (1970)	*Lolium perenne* (perennial ryegrass), *Trifolium repens* (white clover)	80 days (e)	S_W	**G**	
Fresco (1973)	*Galinsoga* spp. (gallant soldier)	8 weeks (g)	W, R_W	**G**	
Landsberg (1974)	*Pyrus malus* (apple)	12 weeks (e)	ΔW	**G**	
Kobayashi (1975)	*Helianthus annuus* (sunflower)	25 days (g)	L_A	**G**	
Wallach (1975)	Natural pasture spp.	10 days (e)	S_W	**R**	
Fukai and Silsbury (1976)	*Trifolium subterraneum* (subterranean clover)	100 days (p)	S_W	**G, L**	Curve also used by Silsbury and Fukai (1977)
Huett and O'Neill (1976)	*Dioscorea* sp. (sweet potato)	39 weeks (p)	W, L_A	**L, G, R, E**	
Hunt and Loomis (1976)	*Nicotiana rustica* (tobacco)	60 days (g)	Tobacco callus W	**G, R**	
Wallach and Gutman (1976)	Natural pasture spp.	215 days (e)	S_W		
Bazzaz and Harper (1977)	*Linum usitatissimum* (flax)	108 days (g)	Leaf N	**G**	
Constable and Gleeson (1977)	*Gossypium hirsutum* (cotton)	180 days (p)	W_S	**G**	
Silsbury and Fukai (1977), Fukai and Silsbury (1977)	*Trifolium subterraneum* (subterranean clover)	26 weeks (p)	S_W	**G**	See Fig. 6.4

Table 6.1 (*Continued*)

Author(s)	Species	Time interval	Primary data	Derived data	Comments
Gray and Morris (1978)	*Lactuca sativa* (lettuce)	8 weeks (p)	W		Harvest date = f (sowing date)
Kimura, Yokoi and Hogetsu (1978)	*Helianthus tuberosus* (sunflower)	18 days (e)	L_A	**G, R**	Individual leaves
Bazzaz and Carlson (1979)	*Ambrosia trifida* (giant ragweed)	50 days (g)	Flower L, Ws		
Bazzaz, Carlson and Harper (1979)	11 tree spp.	160 days (e)	Propagule W		Also used segments, section 7.2
Beale and Thurling (1979)	*Trifolium subterraneum*	16 weeks (e)	Disease % damage	**G, D**	Disease progress curves
Dennett, Elston and Milford (1979)	*Vicia faba* (field bean)	110 days (p)	L_A	**G**	Individual leaves
Marani (1979)	*Gossypium hirsutum* (cotton)	80 days (p)	Boll W	**G**	Time-scale corrected for temperature
Máthé, and Máthé (1979)	*Solanum dulcamara* (woody nightshade)	17 weeks (p)	W, alkaloid content		
Silsbury *et al.* (1979)	*Medicago truncatula* (snail medic)	240 days (g)	W	**G**	
Viragh (1979)	*Quercus cerris* (Turkey oak), *Q. petraea* (sessile oak)	139 days (e)	L_W, L_A		Sun and shade leaves
Wild, Woodhouse and Hopper (1979)	*Lolium perenne* (perennial ryegrass), *Raphanus sativus* (radish)	39 days (g)	**K**		
Dale, Coelho and Gallo (1980)	*Zea mays* (maize)	37 days (e)	**L**		Also used segments, section 7.2

(*Continued*)

Table 6.1 (*Continued*)

Author(s)	Species	Time interval	Primary data	Derived data	Comments
Ferraris and Sinclair (1980)	*Pennisetum purpureum* (elephant grass)	26 weeks (p)	*W*		
Major (1980)	*Zea mays* (maize)	100 days (from anthesis)	Grain *W*	G	

For footnotes see Table 5.1

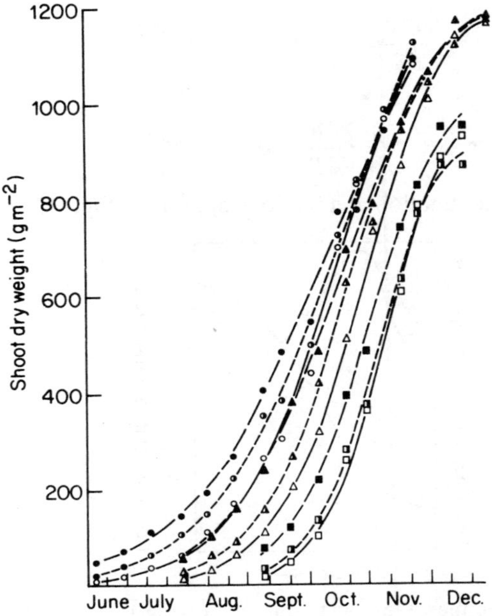

Fig. 6.4 Logistic progressions of shoot dry weight in *Trifolium subterraneum* (subterranean clover) grown in Southern Australia in 1971. Nine combinations of date of sowing and density of planting are shown (data and analyses from Silsbury and Fukai, 1977).

point of inflection occurs at $T = (\log_e b)/c$ but in the W (or $\log_e W$) dimension the curve is asymmetrical, with an inflection at a/e. The linear rearrangements for the purposes of fitting are, for the Gompertz,

$$\log_e[\log_e(a/W)] = \log_e b - cT \tag{6.8a}$$

and for the Gompertz exponential

$$\log_e[\log_e(a/\log_e W)] = \log_e b - cT \quad . \tag{6.8b}$$

In both cases, the double logarithmic transformation is necessary to eliminate the double exponentiation present in equation 6.7. The derivates of the Gompertz are

$$\frac{\mathrm{d}W}{\mathrm{d}T} = abc\mathrm{e}^{-cT - b\mathrm{e}^{-cT}} \tag{6.9a}$$

and

$$\frac{1}{W} \cdot \frac{\mathrm{d}W}{\mathrm{d}T} = bc\mathrm{e}^{-cT} \tag{6.9b}$$

and for the Gompertz exponential

$$\frac{\mathrm{d}W}{\mathrm{d}T} = abc\mathrm{e}^{[a\mathrm{e}^{-b\mathrm{e}^{-cT}} - b\mathrm{e}^{-cT} - cT]} \tag{6.9c}$$

and

$$\frac{1}{W} \cdot \frac{\mathrm{d}W}{\mathrm{d}T} = abc\mathrm{e}^{-cT - b\mathrm{e}^{-cT}} \tag{6.9d}$$

A specimen of the function and its slope appears in Fig. 6.5.

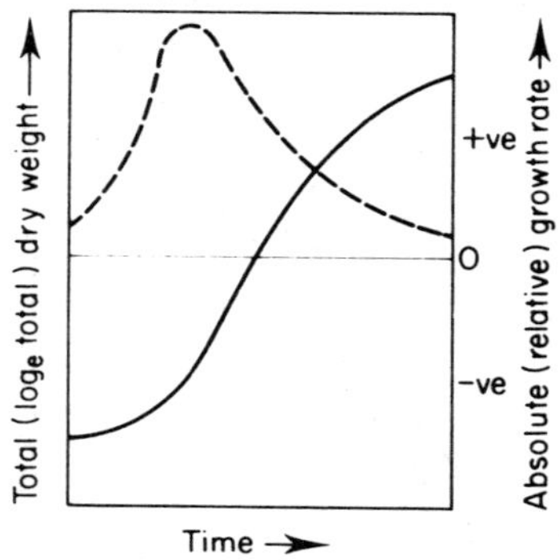

Fig. 6.5 A specimen Gompertz curve showing the progressions of total or $\log_e$ total dry weight (——) and of its slope (– – –).

Lengthy comparative discussions of the properties of the Gompertz curve, especially in relation to those of the logistic, have been given by Winsor (1932) and Richards (1969). The former warned that there seemed to be

> 'no particular reason to expect that the Gompertz curve will show any wider range of fitting power than any other three-constant S-shaped curve. . . . The degree of "skewness" in the Gompertz curve is just as fixed as in the logistic; and it is clear that to introduce a variable degree of skewness into a growth curve will require at least four constants'.

Of the Gompertz and logistic functions, the latter wrote

> 'since many [asymptotic] growth data are characterized by maximal rates somewhere within the range $a/3$ to $a/2$, these can usually be fairly well accommodated by one or other of the two'

though this opinion did not prevent Richards from devising for himself (in 1959) what has since come to be the most important of four-parameter functions (section 6.5). Riffenburgh (1966) and Kidwell, Howard and Laird (1969) have also dealt with the Gompertz function and its properties.

6.4.2 Case studies

The majority of applications of the Gompertz function in plant growth analysis have been connected with the modelling of the growth of individual organs, particularly that of leaves (Table 6.2). Amer and Williams's (1957) treatment of increase in area of individual leaves in *Pelargonium zonale*, of which Fig. 6.6 is an example, showed almost the full span of the function's

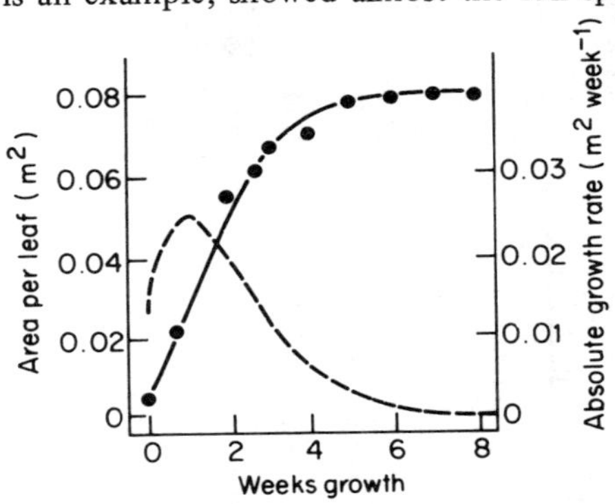

Fig. 6.6 The Gompertizian progression of area in an individual leaf of *Pelargonium zonale* (——), with absolute growth rate (– – –) (data and analysis from Amer and Williams, 1957).

properties within the space of nine harvests, while Hackett and Rawson's (1974) fits to the same variate in *Nicotiana tabacum* (tobacco) demonstrated, like Fig. 6.4, the advantages gained when asymptotic functions appear *en famille* (Fig. 6.7). The latter also rebuts the notion of the 'modular leaf' (section 6.1) just as effectively as the later Fig. 6.13 supports it.

In modelling the growth of whole plants, the Gompertz curve, on the evidence of Kreusler's data from Table 1.1, may well be a reasonable possibility. The skewed progression of slope seen in Fig. 6.5 is a promising gross representation of Kreusler's progression in $\overline{\mathbf{R}}$ (Fig. 2.1b) but, since we currently lack for this function a suitable method of estimating errors of slopes, this precludes any analysis of these data here in a way which is uniform with the other analyses variously performed in this text.

6.5 Richards function

6.5.1 The function and its properties

Unlike the three preceding, this function, proposed by F. J. Richards in 1959, has four-parameters:

$$W\,(\text{or } \log_e W) = a\,(1 \pm e^{(b - cT)})^{-1/d} \tag{6.10}$$

and the second exponent, $-1/d$, involves the additional parameter, d. The term $e^{(b - cT)}$ is a more modern rearrangement of be^{-cT}, which was formerly used in this function in the same way as in the monomolecular, logistic and Gompertz. Derivates of the Richards function are

Table 6.2 Applications of the Gompertz function.

Author(s)	Species	Time interval	Primary data	Derived data	Comments
Amer and Williams (1957)	*Pelargonium zonale* (pelargonium)	8 weeks (e)	L_A	**G**	See Fig. 6.6
Rees (1963), Rees and Chapas (1963)	*Elaeis guineënsis* (oil palm)	2–31 weeks (p)	W, L_A	**R, F, E, L, C**	
Koller, Nyquist and Chorush (1970)	*Glycine max* (soybean)	50 days (from fruiting)	Seed W	**G**	
Koller (1971)	*Glycine max* (soybean)	98 days (p)	Ws	**G**	Component curves added together
Chance and Foerster (1973)	*Persea* sp. (avocado)	83–143 days (after grafting)	Graft H		
Hackett and Rawson (1974)	*Nicotiana tabacum* (tobacco)	20–100 days (p)	L_A	**G, R, E**	Individual leaves, see Fig. 6.7
Arnott (1975)	*Lolium perenne* (perennial ryegrass)	10 days (g)	Ws, lengths		Very early time-scale, component growth
Baker, Horrocks, and Goering (1975)	*Zea mays* (maize)	100 days (p)	L_A		
Constable and Gleeson (1977)	*Gossypium hirsutum* (cotton)	180 days (p)	Ws	**G**	
Pegelow *et al.* (1977)	*Gossypium hirsutum* (cotton)	Relatively short	Hypocotyl length		Full methodological discussion
Christie (1978)	Queensland grassland	10 weeks (e)	W	**G**	

For notes see Table 5.1

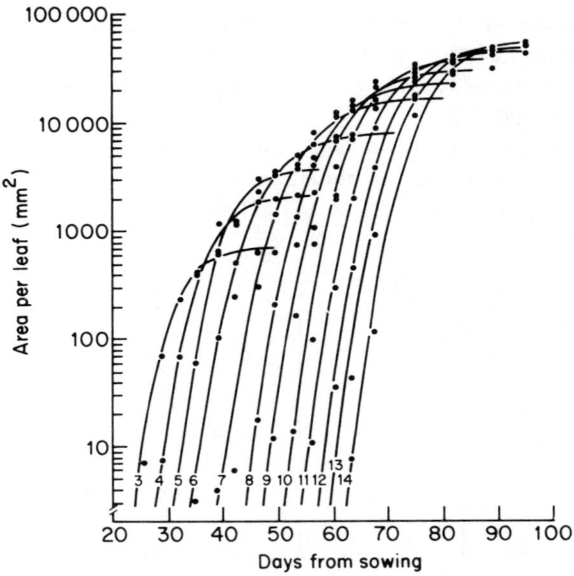

Fig. 6.7 A family of Gompertz curves fitted to the areas of individual leaves (numbered 3 to 14) in tobacco grown in a controlled environment (data and analyses from Hackett and Rawson, 1974).

$$\frac{dW}{dT} = \frac{ace^{b-cT}}{d} \cdot (1 \pm e^{b-cT})^{-(1/d+1)} \tag{6.11a}$$

and

$$\frac{1}{W} \cdot \frac{dW}{dT} = \frac{ce^{b-cT}}{d(1 \pm e^{b-cT})} \tag{6.11b}$$

and for the Richards exponential

$$\frac{dW}{dT} = \frac{ace^{b-cT}}{d} \cdot (1 \pm e^{b-cT})^{-(1/d+1)} \cdot e[a(1 \pm e^{b-cT})^{-1/d}] \tag{6.11c}$$

and

$$\frac{1}{W} \cdot \frac{dW}{dT} = \frac{ace^{b-cT}}{d} \cdot (1 \pm e^{b-cT})^{-(1/d+1)} \ . \tag{6.11d}$$

The previous three functions have had (in equations 6.2, 6.5 and 6.8) linear forms which may be fitted if a value of a is first sought by trial and error. In the case of the Richards function, the originator's (1959) method of fitting was to select both a and d by this method. Subsequently, a long sequence of improvements on this approach has evolved (Nelder, 1961; Causton, 1969; Davies and Ku, 1977; Hadley, 1978), culminating in the most modern method of fitting which enjoys automatically computed starting values, a separate treatment of a and d and the stability of the method over a wide variety of curve types. For complete explanations see Causton and Venus (1981, p. 100).

Two specimens of the function and its slope appear in Fig. 6.8. Four further specimens, displaying various combinations of parameter sign and value, were given by Causton, Elias and Hadley (1978) and Causton and Venus (1981).

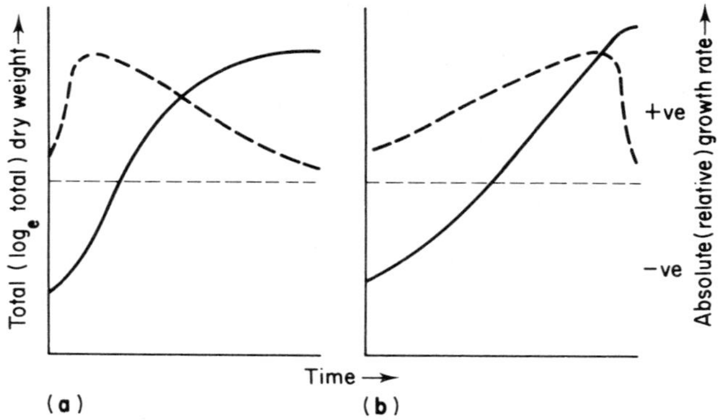

Fig. 6.8 Specimen Richards curves showing the progressions of total or $\log_e$ total dry weight (——) and of its slope (– – –): **(a)** parameter d negative; **(b)** d positive and high.

These latter series were progressions of the form $W = f(T)$, $\log_e W = f(T)$ and $\mathbf{R} = f(W)$; also appearing in presentations uniform with these were single specimens each of the monomolecular, logistic and Gompertz curves, for comparison. Stepwise polynomial exponentials (section 5.6) were compared with the Richards function in the worked examples given by Venus and Causton (1979b).

Causton, Elias and Hadley (1978) pointed out that Richards's rate constant, c, is of particular importance when viewed in combination with d. Thus, when dealing with untransformed primary data, the combination $c/(d + 1)$ is a weighted mean relative growth rate over the whole period, and $ac/[2(d + 2)]$ is the corresponding weighted mean absolute growth rate. When dealing with logarithmically transformed primary data the latter is, of course, a weighted mean *relative* growth rate. The fourth parameter, d, was Richards's reply to Winsor's (1932) plea for a variable point of inflection in asymptotic growth functions (section 6.4). The parameter controls whether or not the function has an inflection, and if so, where it occurs. Figure 6.9, taken from Causton and Venus (1981), shows that with $d = -1$ no inflection is possible; increasing the value of d moves the point of inflection progressively higher up the curve, although this influence operates with gradually diminishing effect.

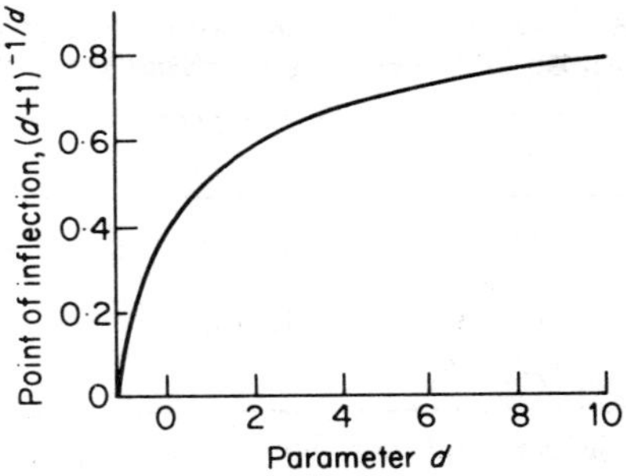

Fig. 6.9 The Richards function: the relationship between parameter d and the point of inflection, expressed as a fraction of the asymptote, a (from Causton and Venus, 1981).

Causton and Venus's work is a lengthy monograph based almost entirely upon the use of the Richards function and should be regarded as the definitive modern source of information for potential users of this curve. Here, we have briefly noted the chief properties of the function; for further information the reader is referred to the aforementioned monograph and to the following selection of the function's extensive methodological literature: Richards (1959), Causton (1969), Richards (1969), Pienaar and Turnbull (1973), Johnson, Sargeant and Allen (1975), Fitzhugh (1976), Savinov, Vasilyev and Schmidt (1977; who repeated Richards's (1959) exposition for the benefit of Russian readers), Causton, Elias and Hadley (1978), Venus and Causton (1979b, c) and Żelawski and Lech (1980).

One final property, which serves to draw together the Richards function on the one hand, and the monomolecular, logistic and Gompertz functions on the other, is the fact that the latter three can each be shown to be special cases of the former (Richards, 1959), or rather, all four can be regarded as variants of a single archetype. This can be written

$$Y = \alpha + \beta\gamma^T \tag{6.12}$$

where Y is again our generalized dependent variate and α, β and γ are special parameters, the particular identities of which vary according to which one of the four growth functions is under consideration (Table 6.3).

Kreusler's data for maize (Table 1.1), appearing for the first time in this chapter because of the aforementioned difficulty in deriving errors of slopes, illustrate, over at least part of their range, one convenient property of the Richards function: that of permitting (through the involvement of four parameters) an inflection not only in the progression itself, but also in its slope (Fig. 6.10, an analysis employing the Richards function, not the Richards exponential, by courtesy of D. R. Causton). Unfortunately, the shape of the early part of Kreusler's progression of untransformed primary data lay only

Table 6.3 Identities of variate Y and of parameters α, β and γ in the generalized asymptotic function $Y = \alpha + \beta\gamma^T$ (after Richards, 1959, 1969).

Function	Y	α	β	γ
Monomolecular	W	a	ab	e^{-c}
Logistic	$1/W$	$1/a$	b/a	e^{-c}
Gompertz	$\log_e W$	$\log_e a$	b	e^{-c}
Richards	W^{1-d}	a^{1-d}	b	e^{-c}

Note: *a, b, c* and *d* are the empirical constants used throughout this chapter and e is the base of natural logarithms.

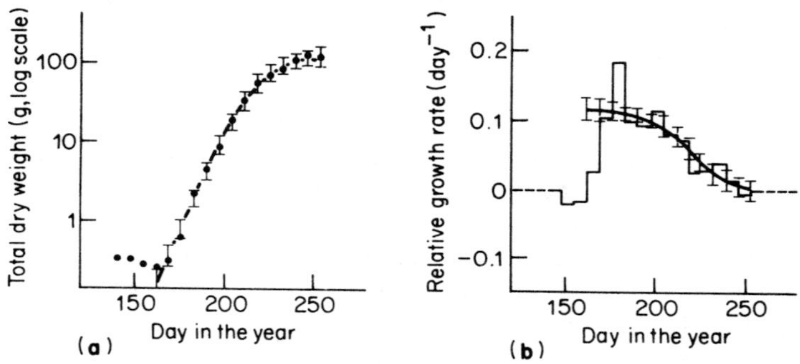

Fig. 6.10 Kreusler's data for maize (Table 1.1) fitted by the Richards function (with the exclusion of the first three data): **(a)** observed values of total dry weight per plant, with the fitted curve and 95% limits; **(b)** fitted values of relative growth rate, with 95% limits, shown against the harvest-interval means.

superficially within the bounds of the Richards function and caused such gross deviations from the fitted curve that the limits about the latter and, particularly about its slope, became unworkably wide. The analysis appearing in Fig. 6.10, therefore, proceeded without the first three primary data.

So, as a model for the growth of whole plants, the Richards function may be circumscribed at both ends; at the beginning, when there is an initial loss in size and at the end, should there be another downturn in the progression due to senescence. Within this range, however, and almost uniformly when dealing with the growth of plant components, its properties are most valuable.

6.5.2 Case studies

The rapidly expanding numbers of applications of the Richards function (Table 6.4) are currently almost equally divided between whole plant and component growth studies. Figure 6.11, from Voldeng and Blackman (1973), shows analyses of dry matter production in cultivars and hybrids of *Zea mays*

Table 6.4 Applications of the Richards function.

Author(s)	Species	Time interval	Primary data	Derived data	Comments
Richards (1959) (and 1969)	*Cucumis melo* (cantaloup melon)	18 days (g)	Hypocotyl L	**G, R**	
Friend, Helson and Fisher (1962)	*Triticum aestivum* (wheat)	50 days (g)	W	**G, R**	
Friend, Helson and Fisher (1965)	*Triticum aestivum* (wheat)	56 days (g)	Ws, L_A	**R, F, E,** LWR	
Causton and Mer (1966)	*Avena sativa* (oat)	70 days (g)	Cell N	**G**	
Causton (1969)	*Acer pseudoplatanus* (sycamore)	24 weeks (g)	W		
Williams and Bouma (1970)	*Trifolium subterraneum* (subterranean clover)	30 days (p)	Lengths, V, FW	**R**	
Pienaar and Turnbull (1973)	*Picea abies* (spruce), *Pinus elliottii* (slash pine)	100 years (p)	H, V, diam.	**G**	
Voldeng and Blackman (1973)	*Zea mays* (maize)	150 days (g)	Ws, L_A	**R**s, **F, E**	See Fig. 6.11
Namkoong and Matzinger (1975)	*Nicotiana tabacum* (tobacco)	To maturity	H	**G**	
Williams (1975)	Several agron. and hort. applications	Various	L, V, FW	**R**s	Early developmental growth, see Fig. 6.13
Woodward (1976)	*Glycine max* (soybean)	95 days (g)	L_A		
Causton, Elias and Hadley (1978)	*Impatiens parviflora* (small-leaved balsam)	40 days (g)	L_A, L_W	**G**s, **R**s	Individual leaves

(*Continued*)

Table 6.4 (*Continued*)

Author(s)	Species	Time interval	Primary data	Derived data	Comments
Dennett, Auld and Elston (1978)	*Vicia faba* (broad bean)	20 days (e)	L_A	**G**	Individual leaves
Nátr, Apel and Kousalová (1978)	*Hordeum vulgare* (barley)	60 days (after flowering)	Kernel, *W*, N, P		
Busey and Myers (1979)	15 turfgrasses and vars.	46(–192) days (g)	*W*	**R**	
Dennett, Elston and Milford (1979)	*Vicia faba* (field bean)	108 days (g)	L_A	**G**	Individual leaves
Littleton *et al.* (1979a)	*Vigna unguiculata* (cowpea)	85 days (g)	L_A	**L, E**	
Littleton *et al.* (1979b)	*Vigna unguiculata* (cowpea)	85 days (g)	Pod *W*	**G**	
Venus and Causton (1979b)	*Helianthus annuus* (sunflower) *Triticum aestivum* (wheat)	36 days (g)	*W*s, L_A	**R**s, ratios, **E**	See Fig. 6.12, cf.s with polynomial exponentials
Causton and Venus (1981)	4 spp.	Various	*W*s, L_A	**R**s, ratios, **E**	See Fig. 6.14
Dennett and Auld (1980)	*Vicia faba* (broad bean)	To leaf no. 8	L_A	**G**	Individual leaves

For footnotes see Table 5.1

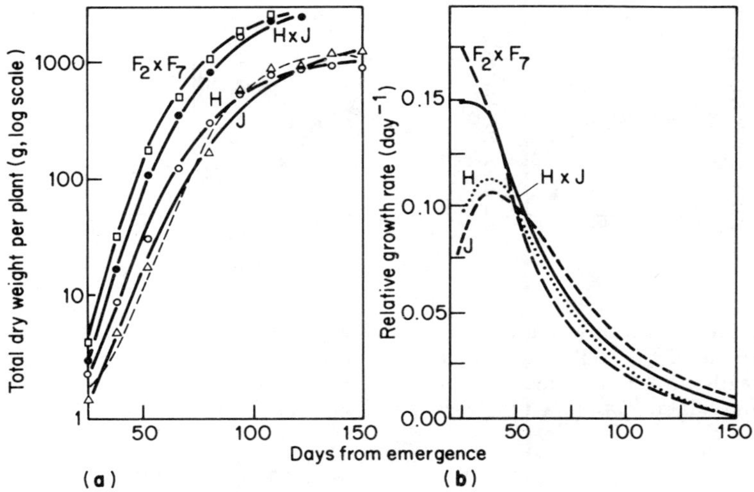

Fig. 6.11 Richards functions fitted to data on the growth of maize, obtained near Oxford in 1965; H and J are two inbred lines and H × J and $F_2 \times F_7$ are their more vigorous hybrids: **(a)** progressions of shoot dry weight; **(b)** derived progressions of relative growth rate (data and analyses from Voldeng and Blackman, 1973).

(maize) which span almost the entire life cycle of the plant and thus fully exploit the properties of the function (section 6.5.1), stopping just short of trouble at the beginnings and at the ends. These examples also reveal, from their final yields per plant, just how far the commercial production of this crop has advanced in the last century (cf. Fig. 2.1 for Kreusler's maize), even though the general form of the progression has not changed. Taking Richards fits to primary data through as far as unit leaf rate, with full statistical treatment, is Venus and Causton's (1979b) analysis of growth in *Helianthus annuus* (sunflower, Fig. 6.12). The longish plateau of more or less constant **R** in this example is clearly due to the antagonistically changing progressions of **F** and **E**. As already mentioned (section 6.5.1), Venus and Causton (1979b) also included in their paper a detailed comparison of this particular analysis with others on the same data done by means of polynomial exponentials (Hunt and Parsons, 1974; section 5.6).

When dealing with patterns of growth in individual plant components, the Richards function has been an indispensible part of several major investigations. In addition to the extensive studies of M. D. Dennett and his co-workers on leaf growth in *Vicia faba* (bean vars.), there has been a long series of observations on the developmental morphology of several species, made by R. F. Williams and his associates. This culminated in the text by Williams (1975), from which Fig. 6.13 is taken. In sheer spectacle this remarkable

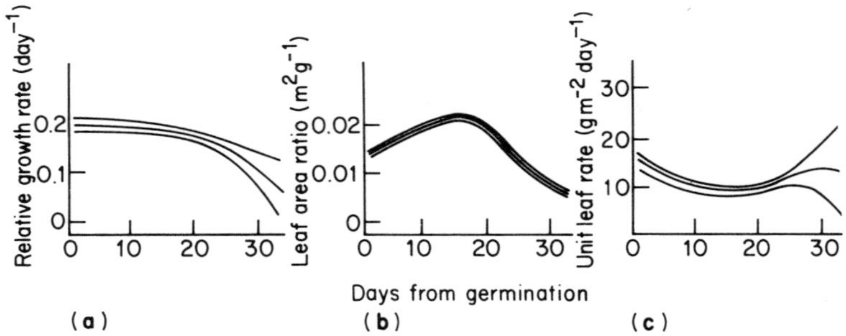

Fig. 6.12 Sunflower grown in a glasshouse at Aberystwyth, showing progressions of **(a)** relative growth rate; **(b)** leaf area ratio and **(c)** unit leaf rate: all with 95% limits and all derived from Richards functions fitted to the primary data, W and L_A (analyses from Venus and Causton, 1979b).

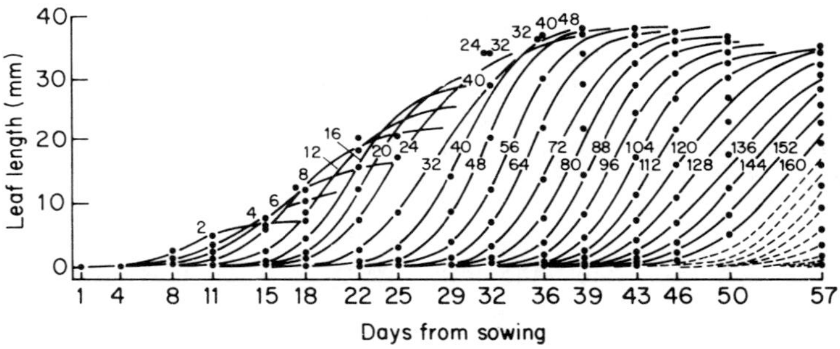

Fig. 6.13 Richards curves fitted to the lengths of individual leaves (numbered 2 to 160) in *Linum usitatissimum* (the cultivated flax; data and analyses from Williams, 1975).

example outdoes even the log-linear display put on by Fig. 5.6 and confirms, if confirmation were needed, that the family context can be exploited to advantage by this function, too. Finally, Causton and Venus's (1981) analysis of the sequential development of area in individual leaves, again in sunflower (Fig. 6.14), shows that a remarkably high initial relative growth rate (almost 1.0 day^{-1}) in the first leaf decays away almost entirely within a fortnight and, throughout, can be followed (within tight limits) by this most flexible of common asymptotic functions.

6.6 Other asymptotic functions

Even the early days of the modelling of the growth of animal populations saw a substantial amount of modification of asymptotic functions in attempts to find one which was uniquely appropriate to this or that particular situation. Mostly, these functions were one variant or another of equation 6.12.

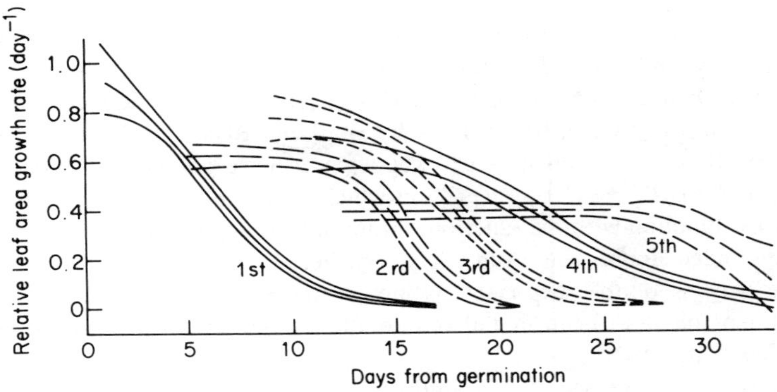

Fig. 6.14 Relative growth rates, with 95% limits, derived from Richards functions fitted to the areas of the first five leaves (youngest on the left) of sunflower grown in a glasshouse at Åberystwyth (data and analyses from Causton and Venus, 1981).

Much of this activity has been reviewed by Melsted and Peck (1977) and Misra (1980).

Meanwhile, as this methodology began to filter through into plant science, more elaborate functions were devised, or were borrowed, in addition to the Richards function (section 6.5). For example, Nelder (1961, 1962) proposed

$$W = a/(1 + c^{-(b + cT)/d})^{d} , \tag{6.13}$$

a development of the logistic (equation 6.4) said to share many of the Richards's properties. This function has been applied by Austin (1964), Austin, Nelder and Berry (1964) and Nichols (1972). Then, the Weibull function,

$$W = a(1 - \mathrm{e}^{-(T/b)^{c}}) , \tag{6.14}$$

has been put forward by Yang, Kozak and Smith (1978) as a flexible sigmoid empirical model for data in forestry, a being the asymptote, and b and c being scale and shape parameters respectively. Campbell, Pennypacker and Madden (1980) used the function to describe the progress of hypocotyl rot in *Phaseolus vulgaris* (snap bean). Lastly, and in contrast to all the foregoing functions in this chapter in which the asymptote a has been expressed in terms of plant size, Żelawski and Lech (1979) have proposed a function with an asymptote at maximum *time*, not maximum size.

Next, we note that many of the old favourites among asymptotic growth curves have themselves undergone a secondary phase of development. The monomolecular function (equation 3.3) was modified by Constable and Rawson (1980) and Rawson, Constable and Howe (1980) in a description of expansion in leaf area in cotton and sunflower respectively:

$$L_{\mathrm{A}} = a + b(1 - \mathrm{e}^{(-cT^{d})}) \tag{6.15}$$

where a is L_A on the day of unfolding, $a + b$ is the maximum (asymptotic) value of L_A and c and d are rate constants, the latter determining the degree of lag in the curve. The logistic function (equation 6.4) underwent development at an early stage, for example by Pearl and Reed (1923) and Richards (1928). Here, its exponent, $-cT$, was replaced by a polynomial function of time, $c_1T + c_2T^2 + c_3T^3 \ldots c_nT^n$, in which $c_1 \ldots c_n$ were the coefficients of this function-within-a-function. More lately, logistic curves have been compounded doubly and even triply together (El Lozy, 1978), in much the same way as double exponentials are created (section 7.6), and the same treatment has also been meted out to the Gompertz (Grunow, Groeneveld and Du Toit, 1980).

Last of all, we see that asymptoticism, complexity and an additional independent variate (section 7.7) in the form of temperature, X, have been fused together by Payendeh, Wallace and MacLeod (1980) into a regression surface describing spore germination in *Entomophthora aphidis*, a Zygomycete pathogenic to aphids:

$$Y = aX^b c^{X^d} (1 - e^{-fT})^{gT^{-h}} \tag{6.16}$$

where e is the base of natural logarithms. Counting the parameters, we find seven: clearly the place at which to bring this chapter to a close.

7

Special approaches

7.1 Introduction

Unlike the preceding two chapters, this one deals in the main not with individual functions used singly but with various combinations of function which can collectively be applied to the same progression of data. Much of this methodology is based upon the polynomial family, so an acquaintance with these functions' properties (Chapter 5) is an important prerequisite for the successful use of the approaches outlined here. Just as the asymptotic functions (Chapter 6) provided possibilities in some ways more elaborate than those of polynomials, so these special approaches provide, in their various ways (and despite being based largely upon polynomials), more elaborate possibilities than do the asymptotic functions. I shall ultimately advance the idea that all of this represents, in the absence of any 'universal growth function', a methodological continuum which the experimenter can enter at a level appropriate to his needs.

7.2 Segments

7.2.1 The segmented principle and its properties

The idea that a single set of data can be fitted by a combination of more than one function, finds its simplest expression in segmented regression analysis. Here, the data set is divided into a series of domains each of which is fitted independently and in isolation by one particular function. The domains may be rigidly demarcated, or they may overlap slightly (though to do so to any great degree would place this approach into the category of the running re-fit, described separately in section 7.3). The extent of each domain, and there may be any number of these, is entirely within the jurisdiction of the experimenter who may reach his decision based either upon statistical or upon biological considerations. For example, a marked change of slope in a progression may provide a convenient point at which to commence a new fit, or periods of growth up to flowering may be considered separately from those beyond, when flowering is believed to initiate a major change in the form of the progression under study.

The technique of segmented regression is valuable, above all, to experimenters who have complex progressions to fit but only simple regressions at their disposal. Naturally, at all stages of the analysis the form of the fitted progression, its derivatives, and the errors of both, are governed by the properties of the function currently holding sway. Whether or not the function provides a happy transition in any of the aforementioned into the domain of its neighbour depends upon the number and variability of primary data in its own and adjacent domains, upon the shrewdness of the experimenter in selecting the extent of each domain, and upon the choice of function itself (in particular, whether or not there is any suggestion of instability due to over-fitting). In short, in each domain there lies the complete responsibility of ensuring a satisfactory treatment of the data that, more generally, overlies all of the functional approach to plant growth analysis.

If the experimenter's aim is to provide seamless derivatives then this inevitably constrains his choice of function. Segmented first-order polynomial exponentials will provide only a histogram-type progression in **R** (like that shown in Fig. 2.1), while second-order polynomial exponentials will produce a zigzag sequence of straight lines. Not until the third order is reached can any semblance of smooth continuity in **R** result from the ensuing second-order curves. Bearing all of this in mind, the experimenter may well ask if segmented regressions are ever likely to be of use to him. To this question the answer must be a heavily qualified 'yes', for provided that all of the circumstances surrounding the analysis of complex progressions with simple instruments are clearly understood, then the substantial advantages of bringing such progressions within the reach of a fitted function will normally outweigh the disadvantages. Of course, would-be Davids will first do best to cast around for superior weaponry before approaching their Goliaths but, this lacking, skill and cunning might yet carry the day.

7.2.2 Case studies

Table 7.1 lists a selection of work in which segmented regressions have been applied. The successful use of many linear segments as extensive 'pseudo-curves' has been demonstrated by Hammond and Kirkham (1949) and Williams (1964) (and, later, in Fig. 7.8), while illustrating the point that functions of different order may be juxtaposed to advantage and provide a felicitous transition in **R**, is the analysis by Hunt and Parsons (1977) of Kreusler's data for maize (Table 1.1). The latter, when compared with the analysis done previously by way of the third-order polynomial exponential (Fig. 5.12), shows both a greatly improved fit and a substantial reduction in 95% limits (Fig. 7.1). On the other hand, it is perfectly possible to find instances in the literature, such as Data and Pratt's (1980) analysis of the development of pod from weight in *Psophocarpus tetragonolobus* (winged bean) in which, on grounds of extent of coverage, representativeness, continuity, choice of domain and heteroscedasticity (Fig. 7.2), a clear victory, alas, rests with Goliath.

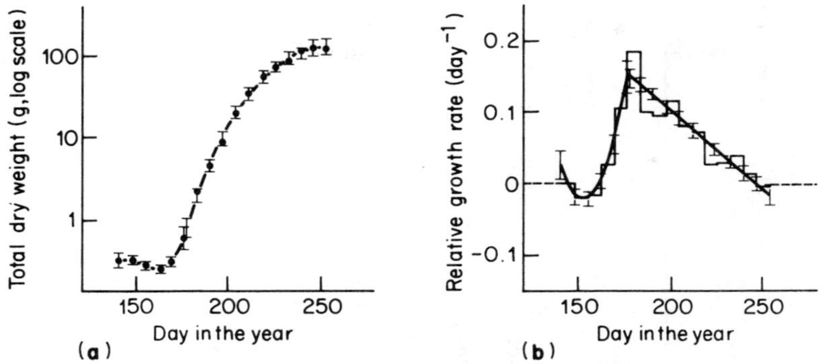

Fig. 7.1 Kreusler's data for maize (Table 1.1) fitted by segmented polynomial exponentials: **(a)** a third-order fit to the period 140–176 days inclusive and a second-order fit to the period 176–253 days inclusive; **(b)** fitted values of relative growth rate, derived from **(a)** and shown against the harvest-interval means. Both parts have 95% limits (analyses from Hunt and Parsons, 1977).

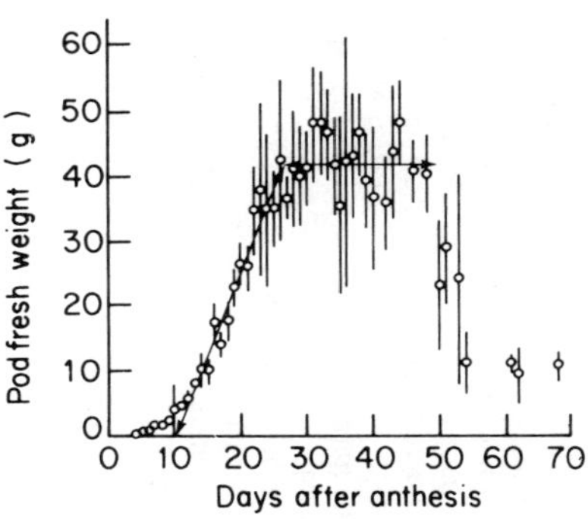

Fig. 7.2 Segmented first-order polynomials fitted to pod fresh weight in winged bean; vertical lines through the data are standard deviations (data and analyses from Data and Pratt, 1980).

7.3 Running re-fits

7.3.1 The re-fit principle and its properties

The running re-fit, otherwise known as the sliding, moving or continuous re-fit or regression, is a logical development of the segmented approach described in section 7.2.1. The data are again divided into sequences of

Table 7.1 Application of segmented regressions.

Author(s)	Species	Time interval(s)	Primary data and order of polynomial, if used	Derived data	Comments
Untransformed primary data					
Maeda (1972)	*Arachis hypogaea* (peanut)	48 days (p)	Leaf *N* (1)	**G**	
Britz *et al.* (1976)	*Ulva lactuca* (sea lettuce)	7 days (e)	Thallus transmittance (2)	**G**	Circadian rhythm
Hall (1977)	*Capsicum annuum* (green pepper)	91 days (after anthesis)	*W* (high-order)	**G**	
Sofield, Wardlaw *et al.* (1977)	*Triticum aestivum* (wheat)	25–55 days (after anthesis)	H_2O content in grain		
Barber (1978)	*Glycine max* (soybean)	120 days (p)	Shoot [N] (1)		
McGreevy (1978)	*Pseudotsuga menziesii* (Douglas fir)	11 years	Trunk diameter (1)		
Klapwijk (1979)	*Lactuca sativa* (lettuce)	200 days (e)	Length of cropping cycle (1)		
MacKinnon (1978)	*Zea mays* (maize, 6 vars.)	120 days (p)	*W*s, **L** (1)		
Dale, Coelho and Gallo (1980)	*Zea mays* (maize)	84 days (e)	**L** (1, logistic)		Temperature-corrected time
Data and Pratt (1980)	*Psophocarpus tetragonolobus* (winged bean)	68 days (after anthesis)	Pod, seed FW, **L** (1)		See Fig. 7.2
Picard, Couchat and Moutonnet (1980)	*Oryza sativa* (rice)	22 days (e)	Transpiration rate		Diurnal variation
Smyth and Dugger (1980)	*Cylindrotheca fusiformis* (a diatom)	70 hours (e)	Cell *N* (1)		

Table 7.1 (*Continued*)

Author(s)	Species	Time interval(s)	Primary data and order of polynomial, if used	Derived data	Comments
Logarithmically-transformed primary data					
Hammond and Kirkham (1949)	*Glycine max* (soybean), *Zea mays* (maize)	120 days (p)	W (1)	**R**	Curves represented by up to 4 segments
Rao and Murty (1968)	*Sorghum bicolor* (sorghum)	142 days (p)	H, W, N (1)	**R**	
Williams (1964)	*Triticum aestivum* (wheat)	25 weeks (p)	W (1)	**R**	Segments cf. with cubics and logistics
Nečas (1974)	*Solanum tuberosum* (potato)	1 hour (e)	FW (1)	– **R**	Logarithmic transpiration curves
Brewster, Bhat and Nye (1975)	*Allium cepa* (onion)	32 days (e)	Ws, R_L, R_W, L_A (1)	**R, E**	
Précsényi *et al.* (1976)	*Zea mays* (maize, 2 vars.)	70 days (p)	W (1)	**R**s, **F, E, C**	
Hunt and Parsons (1977)	*Zea mays* (maize)	113 days (g)	W	**R**	See Fig. 7.1
Cruiziat *et al.* (1980)	*Helianthus annuus* (sunflower)	50 min (e)	ΔH_2O (1)	– **R**	Rehydration kinetics
Fondy and Geiger (1980)	*Beta vulgaris* (sugar beet)	13 hours (e)	^{14}C labelled sucrose (1)	– **R**	Half-lives
Smith and Rogan (1980)	*Agropyron repens* (couch grass)	105 days (p)	Ws, N	**R**	

For footnotes see Table 5.1

domains but the vital difference is that these domains, instead of being merely contiguous, or nearly so, are heavily overlapped, many data points being part of several domains simultaneously. In a sense this procedure is also one variant of the method of 'moving averages' (in which there is a continuous re-calculation of mean values), which may itself be of value to plant growth analysis, especially if the progression under study, though fluctuating, is stationary overall (as in the diurnally based studies of Laval-Martin, Shuch and Edmunds, 1979). Even without stationarity (e.g. in the work of Bernard and Neville, 1978), moving averages, though limited in statistical possibilities, may usefully be employed as simple smoothing functions. However, the method of running re-fit possesses the important enhancement over moving averages in that a sequence not of means, but of fitted regressions (normally polynomials), is applied to the series of domains.

To take a simple example, one general procedure for fitting a running second-order polynomial or polynomial exponential would be to select harvest occasions 1 to 5 inclusive as the first domain, perform a fit, then move to harvest occasions 2 to 6 inclusive, perform another fit, and so on. It is usual to include an odd number of harvest occasions in each domain so that a central value can be taken for comparative work. It is also usual to move each domain only through one harvest occasion at a time, dropping the earliest occasion at each step and adding the occasion immediately succeeding. Ultimately, a curve is built up in which fitted values and derivates arise from the central points only of each of a series of short fitted curves (except, of course, at ends of the whole progression where this strategy is impracticable).

The two matters to which the experimenter must give most attention are the choice of running function and the length of the domain. These may jointly be manipulated to give the required degree of smoothing. In this, it is within the experimenter's power to err in either of two directions. For example, to fit running second-order polynomials to domains only of three points would result in no smoothing at all, while to fit a running first-order polynomial to domains of, say, nine points could well result in too much smoothing, unless the data were such that sequences of nine harvest occasions represented unusually little change in the form of the progression (this would not be the case, for example, in Kreusler's data from Fig. 1.1b).

Methodological background reading in the biological use of running re-fits may be obtained from the publications of DuChateau *et al.* (1972), Thomas, Snyder and Bruce (1977) and, as mentioned previously in the context of orthogonal polynomials (section 5.1), of Erickson (1976).

To illustrate the re-fit principle, a 5-point running analysis has been performed on Kreusler's data for $\bar{W}$ (from Table 1.1). Figure 7.3a shows not a single type of function used throughout, but a mixture of first-, second- or third-order polynomial exponentials, selected for each domain by the stepwise

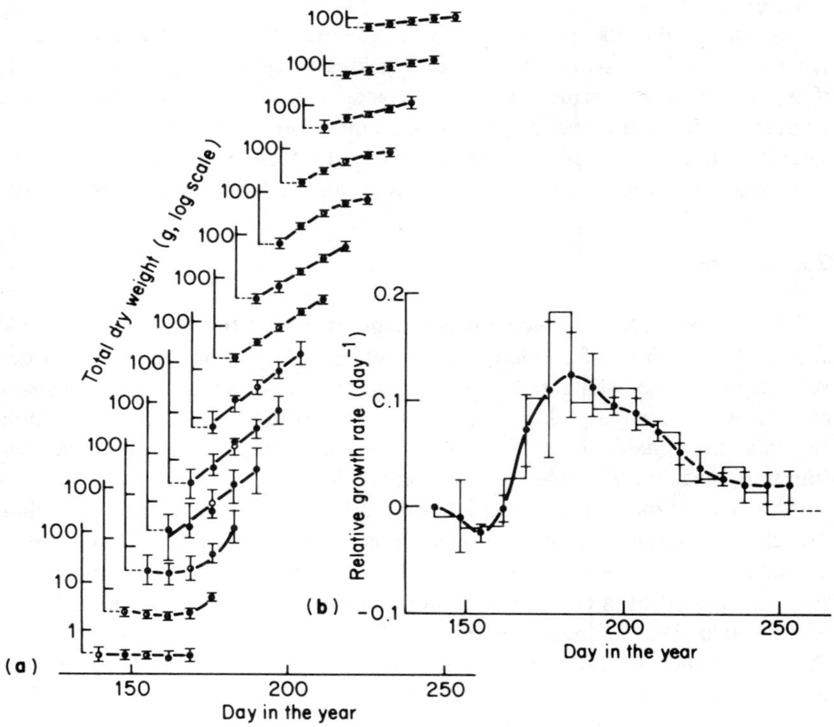

Fig. 7.3 A running re-fit analysis of Kreusler's data for maize (Table 1.1): (**a**) a series of overlapping five-point domains each fitted, with 95% limits, in one of four possible ways by the methods of Hunt and Parsons (1974): no fit (e.g. domain 1), first-order polynomial exponential (e.g. domain 4), second-order polynomial exponential (e.g. domain 3) or third-order polynomial exponential (e.g. domain 2). Points represented by open symbols provided data for (**b**), the progression of relative growth rate derived, except for the first two and the last domains, from the mid-point of each of the re-fits shown in (**a**); harvest-interval means are also shown and 95% limits are included, except at 140 days.

methods of Hunt and Parsons (1974). This ensures that each domain is satisfactorily treated, although the same result could have been achieved by keeping the type of function constant and by varying instead the extents of the individual domains. The resulting progression of **R** (Fig. 7.3b) is easily the most realistic yet encountered in our series, all but the finest detail in $\bar{\mathbf{R}}$ being faithfully represented. Limits, however, are rather wide – an inevitable consequence of the low number of degrees of freedom present in any one domain.

7.3.2 Case studies

One glance at the size of Table 7.2 will show that this powerful and valuable technique has been sadly under-used. For long progressions plentiful in data, if a balance can be struck between choice of function, length of domain, degree of smoothing and degrees of freedom, then at the cost only of considerable, but simple, computational labour the technique can nowadays be executed successfully by even the most modestly equipped laboratory.

7.4 Splines

7.4.1 The principle of splined regression and its properties

Just as the method of running re-fit (section 7.2) can be regarded as a development of the segmented approach (section 7.3), so splined regressions can be interpreted as taking one step still further in the same direction towards the arguable ideal of a seamless and infinitely flexible compounded function, made up of a collection of simpler elements.

The spline function has been likened by Erh (1972) to the more familiar draughtsman's spline, a long, flexible strip used for drawing smooth curves through a set of specified points. In the mathematical spline, the flexible strip is replaced by a chain of separate polynomial functions, each of degree n. Although discrete, neighbouring functions meet at so-called knots. Here they fulfil special continuity conditions, both in the functions themselves and in their first $n-1$ derivates (Wold, 1974).

The root source of this device was the work of Schoenberg (1946); since, it has been developed extensively in the mathematical literature (Ahlberg, Nilson and Walsh, 1967; Greville, 1969). An excellent appraisal of the general value of spline functions in data analysis has been given by Wold (1974) and comparisons with the related technique of running re-fit have been made by DuChateau *et al.* (1972).

The method of splines was first introduced as an alternative to high-order polynomials, or more complicated methods using non-linear functions, where the data to be fitted are obviously beyond the reach of a simple polynomial. Rather than raise the order of polynomial to such a degree that spurious over-fitting took place, with a breakdown in the method of least squares, the original workers in the field considered that a set of low-order polynomials would supply a more effective approach. They encountered two problems: firstly, that of positioning the polynomials so that they would best reflect the underlying trends in the data and, secondly, that of using apparently disjointed and somewhat unrelated polynomials to represent arbitrary domains within the data. The first was solved by creating variables which fixed the domain of each polynomial (not necessarily the same length for each element in the chain) and by using low orders of polynomial so that

Table 7.2 Applications of the running re-fit

Author(s)	Species	Time interval(s)	Primary data	Derived data	Comments
Untransformed primary data					
Erickson and Sax (1956a)	*Zea mays* (maize)	3-hourly, to 80 mm	R_L	**G, R**	
Erickson and Sax (1956b)	*Zea mays* (maize)	3-hourly, to 80 mm	Cell N	**G, R**	
Mengel and Barber (1974)	*Zea mays* (maize)	*c.* 130 days (p)	S_W, %P, R_L	**A**	
Sofield, Evans *et al.* (1977)	*Triticum aestivum* (wheat)	60 days (from anthesis)	Grain W	**G**	Linear phase identified by re-fitting
Logarithmically transformed primary data					
Fisher and Milbourn (1974)	*Brassica oleracea* (Brussels sprout)	28 weeks (p)	W, L_A	**C, L, E**	

For footnotes see Table 5.1.

the shape of each domain within the data could be reflected without over-fitting. The second problem was overcome by constraining the elements at their common end points to agree in position, slope and further derivates, if any. This provided a smooth transition from one polynomial to the next and resulted in a satisfactory overall fit, both mathematically and visually.

Guided by these general requirements, polynomials of varying degree have been splined together; for example, a mixture of straight lines and constants (see Hudson, 1966), several quadratics (Fuller, 1969) or several cubics (Kimball, 1976). In the special application of plant growth studies where logarithmically transformed primary data are involved, the cubic is the simplest element that can be considered. This is because relative growth rate must be free to change smoothly with time, that is, there must be continuity also in the second derivative, rate of change of slope. Splines of lower order cannot possess this feature (see, for example, Fig. 1 of Fuller, 1969).

The principal value of all spline functions lies in their ability to describe lengthy and complicated trends in which 'the particular form of the true function is *not* known; what is known is that the desired function is smooth' (Wahba and Wold, 1975). Splines have far greater flexibility than single regressions (Max and Burkhart, 1976) and provide 'form-free curve fitting' (DuChateau *et al.,* 1972). They can also encompass data which exhibit behaviour in one region that may be unrelated to their behaviour in another' (Kimball, 1976).

Parsons and Hunt (1981) published a spline-fitting program tailor-made for plant growth analysis (that is, providing all of the derivates and limits supplied by the stepwise program of Hunt and Parsons, 1974 (section 5.6.2)). Full details of both of these ALGOL programs are given by Hunt and Parsons (1981) in a brochure which is available on request.

When operating such a program the only decisions left open to the experimenter concern the number and position of knots. Nonetheless, these control the whole nature of the analysis and so correct decisions are essential here if realistic trends in the primary (and, particularly, in the derived) data are to be obtained. For example, consider an analysis of Kreusler's data for maize (Table 1.1), for which progressions of **R** for all numbers of knots from 'none' (the third-order polynomial exponential analysis from Fig. 5.12) to five are presented in Fig. 7.4. Comparisons in work of this type, incidentally, are best made on the derivates, not on the primary data, since this provides a far more sensitive indicator of degree of fit than plots of fitted primary data where, on any reasonable plotted scale, virtually no visual differences would be evident amongst fits by splines employing the higher numbers of knots.

Figure 7.4a shows, as we have previously noted (section 5.4.1), a pronounced lack of fit on the part of the original third-order polynomial

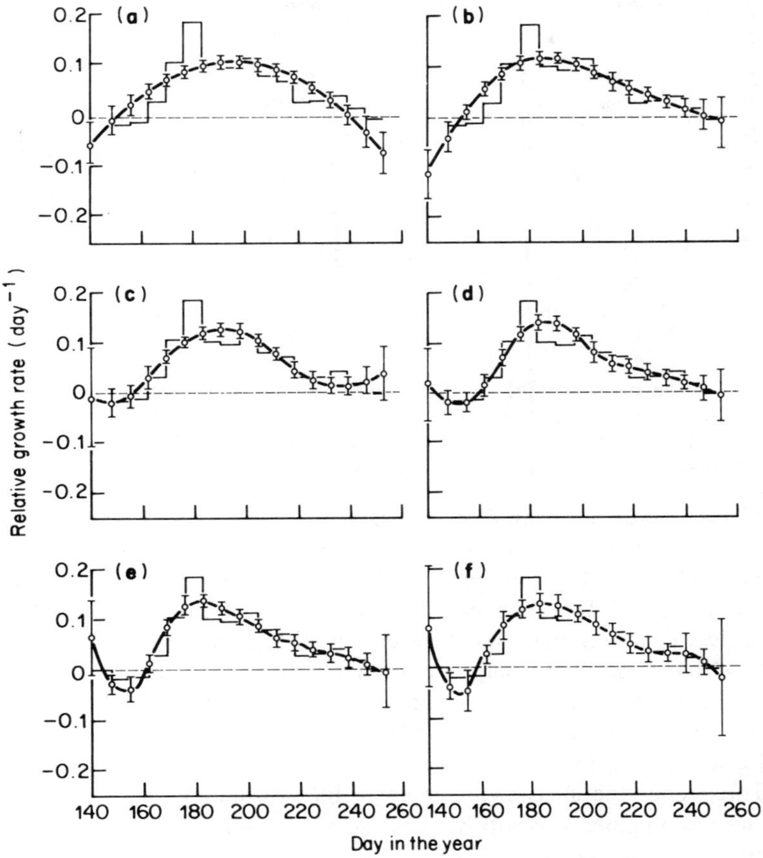

Fig. 7.4 Progressions of relative growth rate, with 96% limits and harvest-interval means, derived from Kreusler's data for maize (Table 1.1). To achieve these progressions, splined cubic polynomial exponentials were fitted to the primary data, employing varying numbers of knots: **(a)** none (a repeat of the analysis shown in Fig. 5.12b); **(b)** one; **(c)** two; **(d)** three; **(e)** four; **(f)** five (analyses from Parsons and Hunt, 1981).

exponential. With one knot in a splined regression (Fig. 7.4b), the 'observed' trend in $\mathbf{\bar{R}}$ is followed in a fairly realistic, but heavily smoothed, fashion (the initial value of $\mathbf{\bar{R}}$, near to zero, is missed). In Fig. 7.4 the repeated changes of slope in the primary data are modelled more successfully by the two-knot spline but spurious values of **R** appear at the end, no doubt due to more important constraints placed upon the splines in other parts of the progression. With three knots, Fig. 7.4d, the approximating function

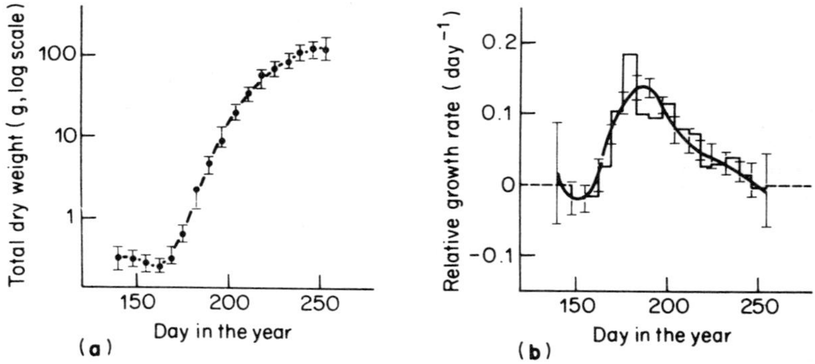

Fig. 7.5 The most suitable analysis selected from Fig. 7.4: **(a)** a three-knot spline fitted to total dry weight per plant in Kreusler's data for maize (Table 1.1), with 95% limits; **(b)** relative growth rates derived from **(a)**, with 95% limits and harvest-interval means (a repeat of Fig. 7.4d).

gains sufficiently in flexibility to follow the macroscopic trend in $\bar{\mathbf{R}}$ in a wholly realistic way. With four and five knots, small-scale detail begins to be pursued by the function but loss of degrees of freedom leads to widening of confidence limits throughout (Figs. 7.4e and f). Even so, these losses are much less damaging than might be expected, due to the fact that adjacent cubic elements agree not only in position but also in their first two derivates. This means that only one degree of freedom is lost with the addition of each knot, despite the fact that this introduces four new parameters into the collection as a whole.

Throughout the analysis presented in Fig. 7.4 the position(s) of each of the declared numbers of knots was decided objectively by a special 'migration' routine (Parsons and Hunt, 1981), though this can be overridden if desired. Based upon the considerations stated above, the three-knot fit may be considered most appropriate, so this is reproduced, with the fitted primary data, in our usual format in Fig. 7.5. This analysis, in fact, is remarkably similar in form to that of the running polynomial exponential, displayed in Fig. 7.3. However, the special advantages gained as a result of the treatment of degrees of freedom in splined regressions have led to the considerably reduced limits evident in Fig. 7.5.

7.4.2 Case studies

Not surprisingly, Table 7.3 confirms that the use of spline functions in plant growth analysis is still in its infancy. For this reason, and because of the great possibilities that the method provides, we need to illustrate applications of this type of function rather more fully than has previously been the case when dealing with others in this text.

Firstly, following the comments made in section 5.6.1, we note that splines may easily be used to satisfy the frequent requirement of making comparisons among a family of related data sets. Hunt and Evans (1980) demonstrated that such regressions can overcome this difficulty, providing biologically realistic families of related curves while retaining sufficient flexibility to ensure a high degree of statistical exactitude in the fits to individual progressions. Figure 7.6 shows progressions of *W* for Badischer früh maize grown in four successive years – data analysed and discussed more fully by Hunt and Evans (1980). One of these curves (for maize grown

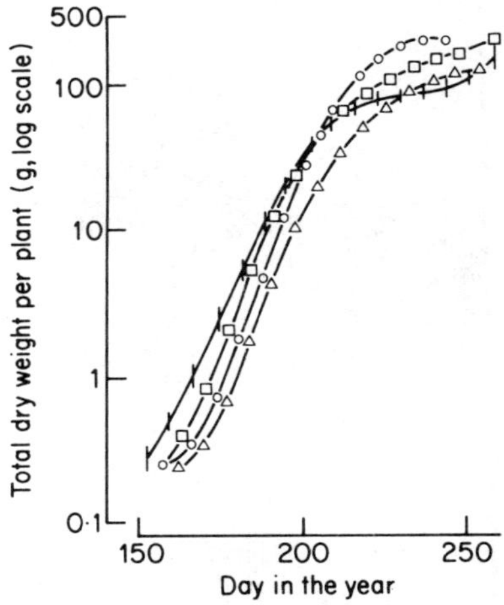

Fig. 7.6 Families of two-knot splines fitted to further data on the growth of 'Badischer früh' maize (not only from 1878 (Table 1.1), but also from the three previous years). The curve bearing 95% limits is that for 1875. Taking these limits as unity the symbols and proportionality of limits for other years are ○, 1876, 1.73; □, 1877, 1.80; △, 1878, 1.55 (data of Kreusler, Prehn and Becker, 1877a, b; and Kreusler, Prehn and Hornberger, 1878, 1879; analyses from Hunt and Evans, 1980).

in 1878) is a slightly truncated two-knot version of the three-knot progression shown in Fig. 7.5a; the others are two-knot splines fitted to data supplied by Kreusler, Prehn and Becker (1877a, b) and Kreusler, Prehn and Hornberger (1878).

Secondly, as an example of a spline curve encompassing a long and detailed progression of data, consider Fig. 8.6, presented later in another context, showing trends in *W* and **R** for *Holcus lanatus* (Yorkshire fog) grown in a productive, controlled environment. This series is an extension of the progression shown in the Fig. 1 of Hunt (1980).

Thirdly, as a more detailed view of events in days 8, 9 and 10 of this long progression for *Holcus*, consider Fig. 7.7. This shows diurnal fluctuations in

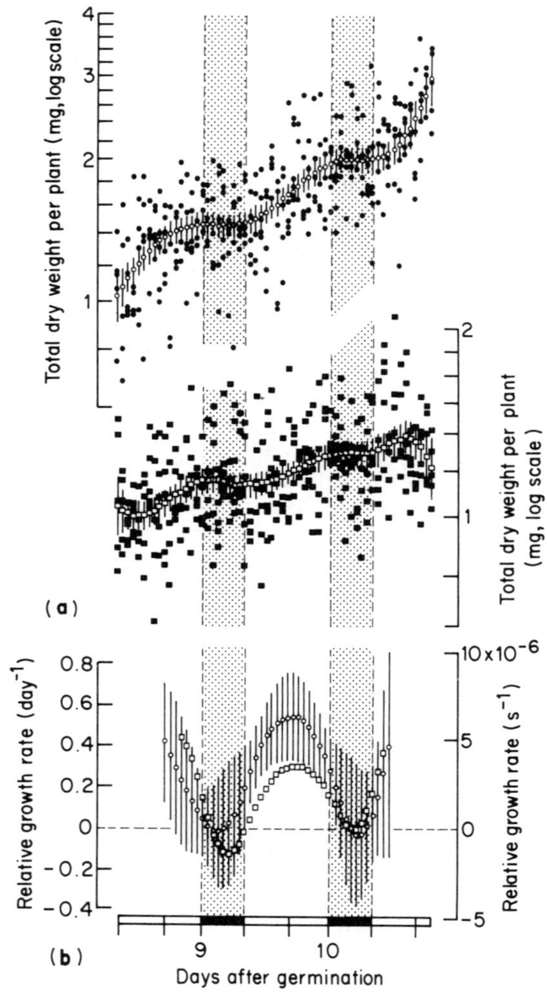

Fig. 7.7 Diurnal patterns of growth in *Holcus lanatus:* **(a)** replicated hourly samples of total dry weight per plant in ●, full-nutrient and ■, low-nutrient treatments, data are fitted by four-knot splines, with 95% limits; **(b)** relative growth rates derived from **(a)**, with 95% limits and dual traditional/SI units (data and analysis from Hunt, 1980).

W fitted by a smooth four-knot spline, which was subsequently used to derive **R** on a more-or-less continuous basis. These data were obtained by way of hourly destructive harvesting by Hunt (1980), who also derived and discussed the corresponding progressions in LWR and ULR.

In all of this, on the number and positioning of knots our experience of

Table 7.3 Applications of spline functions

Author(s)	Species	Time interval(s)	Primary data	Derived data	Comments
Untransformed primary data					
Fuller (1969)	*Zea mays* (maize)	33 years	Grain FW		
Max and Burkhart (1976)	*Pinus taeda* (loblolly pine)	To maturity	Bole diam.		Model of bole geometry
Acevedo *et al.* (1979)	*Sorghum bicolor* (sorghum), *Zea mays* (maize)	16 hours (at 39 or 42 days)	L_L	**G**	Diurnal trends
Logarithmically transformed primary data					
Hunt (1980)	*Holcus lanatus* (Yorkshire fog)	59 hours (at 8 days from g)	W, L_W	**R**s, LWR, $\mathbf{E}_W$	Diurnal trends from hourly destructive harvesting, see Figs 7.7, 7.8, 7.9
Hunt and Evans (1980)	*Zea mays* (maize)	106 days (g)	W, L_A	**R**s, **F**, **E**	Intervarietal and interyear comparisons, see Fig. 7.6
Parsons and Hunt (1981)	*Zea mays* (maize)	113 days (g)	W, L_A	**R**s, **F**, **E**	Data from Table 1.1, see Figs 7.4, 7.5

For footnotes see Table 5.1

the use of spline functions in plant growth analysis has so far led us to support Wold's four main rules of thumb (1974): (1) have as few knots as possible; (2) have not more than one maximum or minimum and one inflection per interval (constraints imposed by the cubic elements themselves); (3) have maxima or minima centred in the intervals; (4) have knots close to inflections.

Next, it may be noted that since the data presented in Fig. 7.7 were obtained from a controlled environment in which the transition from the light to the dark phase was, for the light regime, instantaneous (and for the temperature regime nearly so), it may be more appropriate to acknowledge some discontinuity in **R**, representing the primary data not by seamless splines, but by segments (section 7.2). This is done (using first-order polynomial exponentials) in Fig. 7.8a, where the resulting progression of **R** naturally takes the form of a histogram-type sequence of $\overline{\mathbf{R}}$s (Fig. 7.8b). There is good, but not perfect, agreement of position at each juncture within the primary data (each of which uses one overlapping harvest) and there are wider limits for the dark-phase regressions than for those in the light phase, because of fewer data in the former. As in the splined analysis, Fig. 7.7b, **R** is not significantly negative at night ($P<0.05$).

Finally, we see that near artificial nightfall and daybreak the splines used in Fig. 7.7a are guilty of anticipating forthcoming trends; this may just conceivably be real, but a more likely explanation is over-smoothing because smooth transitions, without discontinuity, are in splines' very nature. In fieldwork this would be no disadvantage at all, but in controlled environments such as the one involved in Fig. 7.7, unless the harvesting is far more frequent even than hourly, some of this defect is inevitable. Thus, the segmented analysis will always possess certain special advantages in cases of this type, in addition to its acknowledged disadvantages.

However, if we believe that, even in the mechanically 'saw-toothed' environment of the growth cabinet, a truly discontinuous progression in **R** does not occur (it may merely undergo transitions too rapid to be followed by normal methods of access), it may be that, given more frequent and elaborate destructive or gasometric observations, spline functions may again be brought into play as mathematical representations of empirical data. It is possible to predict what kind of progression in **R** may ultimately lie behind growth in the saw-toothed environment, as the sketch presented in Fig. 7.9 shows. This draws on the progression in **R** suggested (for a natural environment) by Causton (1977, Fig. 11.2) and makes use of the following assumptions:

Fig. 7.8 (*opposite*) **(a)** A diurnal progression in total dry weight per plant in *Holcus lanatus* (full-nutrient data from Fig. 7.8) fitted by four segmented first-order polynomial exponentials (section 7.2), with 95% limits, in domains which overlap by one harvest occasion; **(b)** relative growth rates derived from **(a)**, with 95% limits (data from Hunt, 1980).

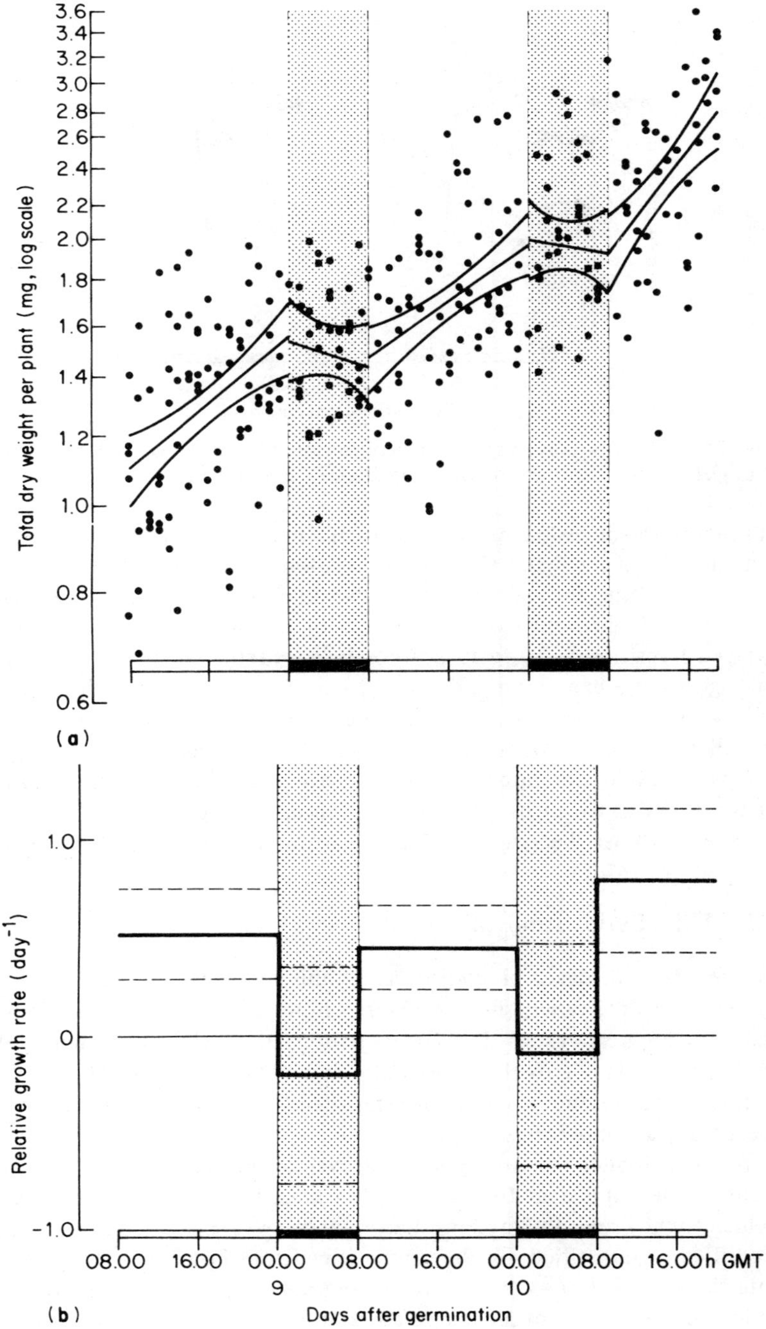
3.6
3.4
3.2
3.0
2.8
2.6
2.4
2.2
2.0
1.8
1.6
1.4
1.2
1.0
0.8
0.6
Total dry weight per plant (mg, log scale)
(a)
1.0
0
-1.0
Relative growth rate (day⁻¹)
08.00
16.00
00.00
08.00
16.00
00.00
08.00
16.00 h GMT
9
10
Days after germination
(b)

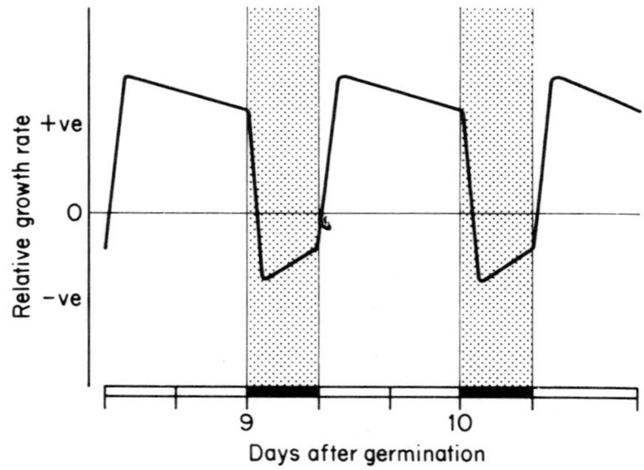

Fig. 7.9 The stylized, predicted progression of relative growth rate which is thought to underly those depicted in Figs 7.7b and 7.8b.

(1) there is a somewhat delayed rise to maximum light-phase **R** following the switching on of the lights; (2) there is short-term fatigue in the photosynthetic apparatus during the light phase, leading to a downward drift in **R** towards nightfall (this drift may or may not be of this linear form); (3) there is a delayed transition to a negative carbon balance following the switching off of the lights (here exaggerated, like (1), to around 2 hours); (4) there is a progressive loss of negativity in **R** as the night wears on, due to respiratory 'stabilization' (as in Evans, 1972, Fig. 12.13); (5) finally, there is a repeat of the whole cycle, ultimately with progressive daily changes in the magnitude of the peaks and troughs resulting from ontogenetic drift on a life-long scale.

When such data become available, the instrument of their analysis lies ready.

7.5 Time Series Analysis

In everyday language, of course, all of this book may be said to be concerned with time series, since it almost wholly deals with series of plant data and their progressions on time (Introduction, p. 3). However, the functional approach to plant growth analysis also impinges, at its upper limit of complexity, onto another separate and fast-developing field of activity, known specifically as Time Series Analysis.

Suitable fodder for this type of analysis is typically a series of measurements made with great frequency and at great length. These may be full of hidden trends and fluctuations that cannot even be guessed at, let alone identified or quantified, by the simple expedient of plotting-out the primary data themselves. For example, take a continuous series of, say, hourly observations or recordings of air temperature at one outdoor location. Here, there will be diurnal fluctuations, periods of favourable and unfavourable weather, seasonal trends, good and bad years and long-term climatic drift, all superimposed in an hierarchical series of pattern upon the hourly primary data. In cases such as these, it is the task of Time Series Analysis to unravel the underlying levels of stationarity, lengths of seasonality, frequencies, amplitudes and phases that lie within the data. To do this satisfactorily can require a minimum of perhaps ten data points *per feature*.

The method chosen to execute a Time Series Analysis may either be time-based (autocorrelative) or frequency-based (spectral). Some of the autocorrelative approaches bear similarities to the methods of moving averages and running re-fit (section 7.3), but many other variants exist, too. A major exposition of the special methodology of Time Series Analysis has been made by Kendall (1973). Shorter, more introductory, accounts have been provided by Anderson (1976) and Makridakis (1976). The latter is particularly broad and approachable, but is also subject to the comments of Anderson (1977).

Meteorology, economics and industry have most frequently claimed applications of Time Series methods. In biological sciences, there are reviews of the technique in connection with the analysis of the human growth curve (Hirschfeld, 1970a, b) and of microbiological populations (Grimm, 1977). The only botanical application I have found has been in the work of Chow and Tan (1979). These authors followed the yield of latex from trees of *Hevea* sp. (rubber) and were able to fit a Time Series that included relatively short-term fluctuations of yield within individual cycles of tapping/recovery, upon which was superimposed a broader trend stretching over many tens of cycles (Fig. 7.10). There were also substantial differences noted between alternative tapping systems and alternative systems of weed control by herbicides.

Time Series Analysis overlaps to some extent with the upper reaches of

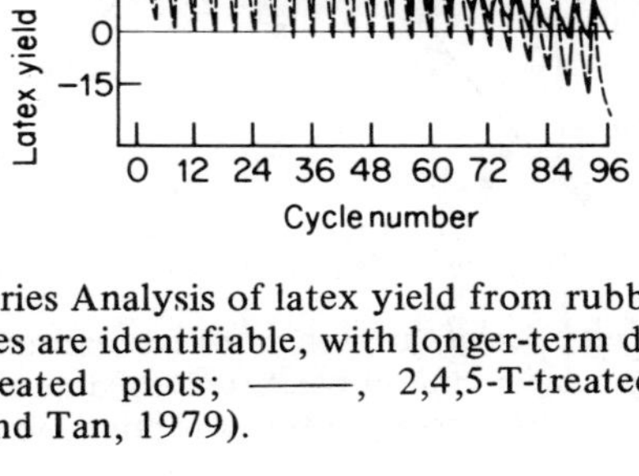

Fig. 7.10 A Time Series Analysis of latex yield from rubber trees. Individual tapping/recovery cycles are identifiable, with longer-term drift super-imposed; – – – –, ethephon-treated plots; ——, 2,4,5-T-treated plots (data and analysis from Chow and Tan, 1979).

splined regression analysis and further instances may yet emerge of very long, and almost inevitably non-destructive, series of measurements (such as automatically recorded girths of tree boles) which may benefit from one or more of the particular advantages of this analytical approach.

7.6 Other special functions

This section deals with a selection of special and relatively infrequently used functions. The reader should be prepared to proceed to the sources cited for any which promise to be of particular value since, alas, there is space here to do little more than to note that they exist. The generalized notation (see section 2.1) is used throughout.

We saw, in the work of Huzulák and Matejka (1980; see Table 5.3), how the third-order polynomial could be used to approximate to a sine function with a diurnal harmonic periodicity. It is quite open to the experimenter, of course, to use the genuine article:

$$Y = a + b \sin(\phi X), \tag{7.1}$$

where a and b are constants, ϕ is 2π radians/the period of one cycle (in this case $360°/24$ hours) and X is hours from the start of the cycle. The special features here are that the progression is bell-shaped and totally symmetrical, but there must be no doubt as to the length of each cycle and this must be specified in advance. Such a curve, on an annual basis, is illustrated in Fig. 7.11, using the data on crop growth assembled by Scotter, Clothier and Turner (1979).

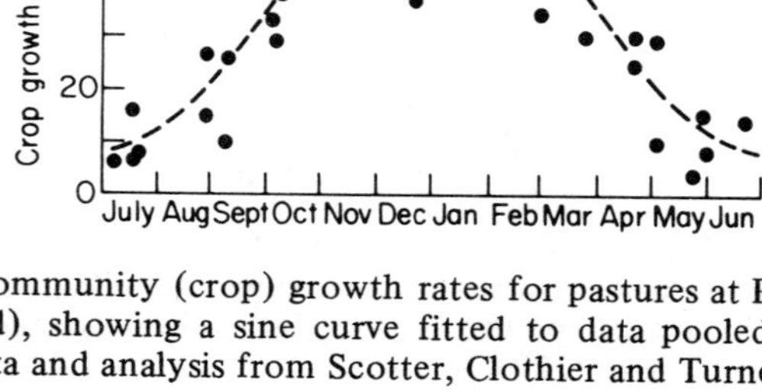

Fig. 7.11 Community (crop) growth rates for pastures at Palmerston North (New Zealand), showing a sine curve fitted to data pooled over the period 1974–78 (data and analysis from Scotter, Clothier and Turner, 1979).

A long-standing elaboration of the sine transformation is the Fourier transformation, for

> 'In 1807, Joseph Fourier showed that any time series . . . can be expanded into a sum of sine and cosine terms which can approximate the series . . . as close as one wishes, provided that the number of sine/cosine terms is large enough' (Makridakis, 1976).

This, of course, is a process analogous to the ultimate over-fitting that is possible when using polynomials (section 4.6). Again, the number of terms used should be restricted in order to secure some degree of smoothing. Kimball (1974) has discussed such uses of Fourier transformations and Gandar (1980) and Steer and Blackwood (1978) have both employed the transformation, and derivates from it, the latter using it in its simplest form (i.e. with most smoothing):

$$Y = a + b \cos (\phi X) + c \sin (\phi X) \ . \tag{7.2}$$

The so-called double exponential equation has been used by Ryle, Cobby and Powell (1976) and Jones (1979). It is the sum of two exponential curves

$$Y = a + b\mathrm{e}^{-cX} + d\mathrm{e}^{-fX} \tag{7.3}$$

where a is an intercept term and e is the base of natural logarithms. Parameters b and c control the response of Y to X within one set of controlling conditions, and d and f the response within another set or within a subsequent range of the same set, the two sets acting additively. Unless $b = d$ and $c = f$ the resulting curve is a hybrid between two different exponential curves, sharing some of the properties of both, as the Fig. 2 of Jones (1979) demonstrates.

Nelder (1961) outlined the utility of a family of curves which he termed inverse polynomials of which the

$$Y = X/(a + bX + cX^2) \tag{7.4}$$

used by Bunting (1972) and Cock *et al.* (1979) is a simple example (all parameters being arbitrary constants). These curves all pass through the origin.

Hammerton and Stone (1966), in the course of a study in which they compared classical derivates (sections 2.2, 2.3), derivates obtained by classical methods from values taken from fitted curves (section 4.4) and truly functional derivates, employed what was virtually a second-order polynomial exponential:

$$Y = a + b^{(cX - dX^2)} . \tag{7.5}$$

They fully discussed the properties of their function and provided notes on the fitting procedure and the evaluation of the degree of fit obtained. They likened the function to the aforementioned inverse polynomials.

Another four-parameter function was used by Humphries (1968). The form

$$Y = a + (b + cX)d^X \tag{7.6}$$

produced asymmetrical bell-shaped progressions, where a, b, c and d are arbitrary constants. Still another four-parameter function, illustrating the principle that numbers mean little when form can be so plastic, was employed by Wild, Woodhouse and Hopper (1979). This was

$$Y = a + b/(1 + \mathrm{e}^{c(X-d)}) , \tag{7.7}$$

where e is the base of natural logarithms. In truth, the progressions of this form which were actually illustrated by these authors hardly justified such mathematical complexity.

Finally, we note what should by now have become obvious: that with modern computing software, experimenters now have the opportunity of juggling with parameter numbers and mathematical structure to produce functions 'tailor-made' to their own particular purpose. These can then be fitted to their data by such methods as maximum neighbourhood (Marquardt, 1963). For example, a five-parameter function of this type was developed by Preece and Baines (1978) to describe the highly idiosyncratic, 'stop-go' progression on time, not of plant data, but of the human growth curve for height:

$${}_{\mathrm{X}}H = {}_{1}H - \frac{2({}_{1}H - {}_{\Theta}H)}{\exp[a(X-\Theta)] + \exp[b(X-\Theta)]} \tag{7.8}$$

where ${}_{\mathrm{X}}H$ is height at time X, ${}_{1}H$ is final adult height, a and b are rate constants, Θ is a time constant and ${}_{\Theta}H$ is height at $X = \Theta$. Naturally, the more specific such functions become the more they lose in generality and there are often difficulties, also in deriving errors for fitted values predicted by the curve, or for first or second derivates of the curve.

7.7 Response surfaces involving additional independent variables or variates

We saw in section 2.4 how it could be advantageous, on occasion, to plot, or even fit, plant data or derivates as a function of some single independent variate other than time. Many possibilities exist in that direction. Here, we briefly consider the ways in which independent variables or variates in addition to time can be incorporated, not into single approximating *curves*, but into multi-dimensional approximating *surfaces*. The great advantage gained in this approach is that such independent variables or variates, often meteorological, become not merely external conditions subject to the usual paired comparisons of time series, but quantities fully integrated into the analysis and capable of providing all of the quantitative appreciation of interpretation and control which we enjoy from time itself.

The usual way of achieving this goal has been to incorporate these additional quantities into a multiple regression equation (see Yarranton, 1969; *caveat* Mead, 1971; and see Mead and Pike, 1975 and Terman and Nelson, 1976), employing computational procedures such as the program BMD02R of Dixon (1973). In its simplest form such a regression, explaining the dependence of total dry weight per plant jointly upon some environmental variate (E) and time, would be

$$W\,(\text{or } \log_e W) = a + bE + cT \quad . \tag{7.8}$$

Regression surfaces of this form can be elaborated in any or all of three ways: (1) additional independent variates may be added, E_1, E_2 . . . (a total of any more than two preclude the possibility of plotting the surface on a simple '3-dimensional' diagram); (2) non-linearity may be introduced into any or all of the terms, T^2, T^3 . . .; (3) interactions between terms can be included, ET, ET^2 . . . .

As examples, see the relatively simple surface fitted by Mahmoud *et al.* (1981) to $\log_e W$ versus time and temperature in seedlings of two *Acacia* spp. grown in a series of controlled environment experiments (Fig. 7.12),

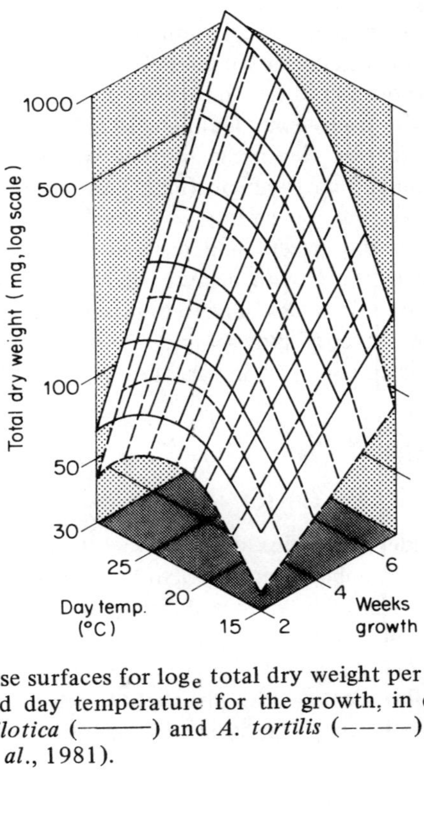

Fig. 7.12 Response surfaces for $\log_e$ total dry weight per plant as a function of seedling age and day temperature for the growth, in controlled environment, of *Acacia nilotica* (———) and *A. tortilis* (– – – –) (data and analyses from Mahmoud, *et al.*, 1981).

and the equation of the relatively complex surface (employing all of the three forms of elaboration listed above) which was used to represent the level of photosynthetic activity in *Calluna vulgaris* (heather) by Grace and Woolhouse (1970):

$$P = (13.7L - 11.0L^2 - 0.4)T - (0.32L + 0.30L^2 - 0.2)T^2 + (0.01L + 1.15L^2 + 0.06)H^2 - (1.36L + 2.0L^2 + 0.2)A + (0.002L - 0.0031L^2 - 0.0003)A^2 - (67.3L + 176L^2 - 1.062)F + (160L - 308L^2 - 17) . \qquad (7.9)$$

Here (and here alone), P is photosynthetic rate, L is light intensity, T is temperature, H is temperature prehistory, A is leaf age, and F is a flowering index (details of the units and origins of each of these dimensions are *loc. cit.*).

Other applications are listed in Table 7.4. Strictly speaking, not all of these display the straightforward incorporation of additional quantities into a single response surface involving time (e.g. Brougham, 1955; Glenday, 1955, 1959; Erdös, 1980) but, nonetheless, the effect of these various approaches has been broadly the same as if they had done so.

Notwithstanding the fact that the great majority of applications in this field have used the statistically linear methods of multiple regression analysis, it is possible to find instances of non-linear functions into which environmental variables in addition to time have been incorporated. Examples of this have been provided by Wallach and Gutman (1976), and by Barnes (1977).

Finally, we consider an elegant analysis by Warren Wilson (1981) which not only linked time and environment (light flux density) together as determinants of plant size, but did so in a way which also brought together the concepts involved in the growth and analysis of individuals (section 2.2) and those involved in the growth analysis of populations and communities (section 2.3). Warren Wilson's 'universal' expression took the form of a detailed subdivision of crop growth rate, notated as the product of the number of plants per unit area, N, and absolute growth rate per plant:

$$N \cdot \frac{\mathrm{d}W}{\mathrm{d}T} = I_0 \times \frac{N}{I_0} \times W \times \frac{L_W}{W} \times \frac{L_A}{L_W} \times \frac{J}{L_A} \times \frac{1}{J} \cdot \frac{\mathrm{d}W}{\mathrm{d}T} \qquad (7.10)$$

(a) (b) (c) (d) (e) (f) (g)

where I_0 is the incident light flux density and J the light flux intercepted by the plant. The terms numbered on the right-hand side of equation 7.10 were grouped by Warren Wilson in various different ways, each providing a subdivision of crop growth rate appropriate to one particular stage of crop growth: (1) taking (a), the product of (b) – (f) and (g) respectively obtained the incident light flux density, the plant's 'intercepting efficiency' and the plant's 'utilizating efficiency' (a 'type (iii)' quantity, section 2.1); (2) the

Table 7.4 Applications of multiple linear regressions involving independent variables or variates in addition to time.

Author(s)	Species	Time interval	Primary data	Independent variates	Comments
Untransformed primary data					
Brougham (1955)	3 pasture spp.	84 days (e)	W	T, temperature, rainfall	Weather effects subtracted frpm plant data, see equation 7.9
Grace and Woolhouse (1970)	*Calluna vulgaris* (heather)	4 hours (e)	Photosynthetic rate	T, plus 4 environmental variates	
Fraley and Whicker (1973a, b)	Prairie community	3 years	Species N	T, γ-irradiation	
Barnes (1977)	6 spp., mainly crops	Various	**G**	T, Ws	
Pearson, Aldwinckle and Seem (1977)	*Gymnosporangium juniperi-virginianae* (cedar apple rust fungus)	6 hours (e)	% teliospore germination	T, temperature	
Thomas and Norris (1977)	*Lolium perenne* (perennial ryegrass)	*c.* 32 weeks (p)	L_L	T, soil temperature, insolation	
Elliott and Peirson (1980)	*Phaseolus vulgaris* (bush bean)	6 days (e)	L_W, nitrate reductase	T, CHCA, 2,4-D	See footnote
Erdös (1980)	*Zea mays* (maize)	95 years	Grain FW	'Weather, agrotechnics, soil'	
Logarithmically transformed primary data					
Abul-Fatih, Bazzaz and Hunt (1979)	*Ambrosia trifida* (giant ragweed)	61 days (e)	W, L_A, **R**s **F**, **E**	T, planting density	

(Continued)

Table 7.4 (*Continued*)

Author(s)	Species	Time interval	Primary data	Independent variates	Comments
Singh *et al.* (1979)	*Corchorus capsularis* (jute)	19 weeks (p)	W	Plant density, H, basal diam.	Pre-harvest forecasting
Mahmoud *et al.* (1981)	*Acacia nilotica*, *Acacia tortilis* (Sudanese acacias)	49 days (p)	W, ratios, H, $\mathbf{B}$	T, temperature	See Fig. 7.12

CHCA is cyclohexanecarboxylic acid concentration
2,4-D is 2,4-dichlorophenoxyacetic acid concentration
For other notes see Table 5.1

products of (a) and (b) and of (c) – (g) represented plant density and absolute growth rate; (3) the products of (a) – (c) and of (d) – (g) represented biomass and relative growth rate; and (4) the products of (a) – (e) and of (f) – (g) were leaf area index and unit leaf rate.

Though lacking, as yet, any detailed evaluation, this analysis shows great promise. This is particularly so since smooth progressions of the six variates, once obtained, would enable equation 7.10 to be solved on an instantaneous basis and provide seamless progressions in its various terms, progressions capable of spanning the entire life cycle of the plant from seed to seed.

8

Finalé

8.1 Which growth function?

'. . . in adapting the word [growth] to scientific purposes authors have tended to frame their own definitions. This is both an attractive and a dangerous process: attractive, in that scientific usage requires a precise statement, translatable into terms of quantities which can be measured; and dangerous because we are trying to define natural phenomena in terms of ideas which are products of the human mind. Experience shows that a definition so framed can easily become a Procrustean bed into which nature must be fitted either by stretching it or lopping a bit off.'
G. C. Evans (1972)

8.1.1 In general

The problems surrounding the selection of words to define the concept of growth apply with equal force when mathematical functions are selected to describe its actuality. Evans's delineation thus extends well beyond its ostensible context into one where the experimenter's perception of the underlying reality of growth can either be helped or hindered by the combination of biologically generated observational data and intellectually generated mathematical functions. If there is no harmony between these two, what ultimately enters the experimenter's mind is a corrupt (that is, false or unrepeatable) perception of the phenomenon under study. To avoid this, it would help to survey the functions in a single sweep, looking for guidelines to correct selection.

The preceding three chapters have outlined, under the various main subheadings, thirteen instruments for executing the functional approach to plant growth analysis. These methods ranged from the simplest of those currently available and in use (the first-order polynomial, section 5.2), to the most complex (Time Series Analysis, section 7.6). In addition, Chapters 6 and 7 each included a 'rag-bag' which listed other functions of special, but minor, interest.

Now, we saw in Chapter 3 how the search for a single and universal plant growth function, the parameters of which form some simple and direct mechanistic analogue of the workings of the plant, has long been abandoned (at least in respect of the study of the whole plant over a sizeable portion of its life cycle) since the partial successes previously gained in this direction in animal science have proved to be a poor influence on those studying the very different world of plant growth and development. So, what remains now, and continues to develop, is not the search for this philosopher's stone but an extensive series of empirical methodology. Here and there this may directly yield some mechanistic insight of its own into plant processes but normally it finds itself supported by a rationale no more elaborate than that it offers an enhanced representation of scientific reality (section 3.3). This rationale is limited, but it is sound and none can offer more. Where, then to start?

Broadly, Chapters 5, 6, and 7 formed a series of increasing complexity, both with respect to each other and within themselves. Hunt (1981) skimmed this field and put forward a rough-and-ready guide, identifying the point of entry into this continuum which was most likely to lead to a satisfactory analysis. This was based largely upon the number of data or harvest occasions that were available. The lowest order possible of polynomial or polynomial exponential was suggested for series of up to about six observations. Non-linear functions, especially the Richards, were suggested as requiring upwards of six observations, even neglecting their asymptotic properties, and the more sophisticated methods such as splined regression were held to be useful only upwards of about twelve observations. These upper limits might be raised if trends, although lengthy, were unusually simple in form. There was held to be no upper limit, either in length or in complexity, to the usefulness of the splined regression, although the techniques of Time Series Analysis would sooner or later come into play when faced with multitudinous data in complex progressions.

This simplified scheme may still be of preliminary value but, of course, in dealing so markedly with number it takes too little account of form. Hence, frequent measurements on the growth of unicellular organisms or young plants or parts of plants in certain conditions may, though numerous, call for no more than the first-order polynomial exponential; while a progression numbering a mere handful of data, but recorded against the background of a violently changing environment, may contain such near-discontinuities as to justify, or at least be forced into employing, a splined regression (e.g. those data of Rorison, Peterkin and Clarkson, 1982). Superimposed upon the questions of number and form or complexity there is also the question of asymptoticism. Here, number is certainly important, for there would be little point in fitting, say, four data points by any of the functions discussed in Chapter 6, even were the progression unmistakably asymptotic with perhaps the two or three last values lying on a plateau. Similarly, we have already noted (section 6.1) how fitting asymptotic functions to (numerous) non-asymptotic data, or vice-versa, plays upon the weaknesses, not the strengths, of the functions concerned.

Our methodological pathway then, does not proceed in a simple fashion through Chapters 5, 6 and 7 in that order. Rather, it can be fancied as a way which begins: first-, second-, third-order polynomial (with due deference *en route* to the advantages and disadvantages of the stepwise principle), and then proliferates into a choice of more complex byways: the high-order polynomial, ultimately a dead end but possibly serviceable in its proximal reaches; segments, rough but practicable; running re-fit, smooth but costly; splines, easy enough and unlimited in extent, but requiring the right computational vehicle; and Time Series Analysis, ditto, but also remote and untried. Midway along this journey the parallel approaches of the three- and four-parameter asymptotic functions present themselves. These, almost by definition, will lead to a full stop. If a tidy termination is what is desired, then none other will do; if not, then they will naturally impose their own harmful limitations. Of these ways, the Richards is by far the most accommodating, but at the cost of an additional parameter.

8.1.2 For Kreusler's data

Kreusler's (1879) data on the progression of total dry weight per plant in 'Badischer früh' maize grown in 1878 (Table 1.1) have thus far appeared analysed in no less than ten different ways in this text:

Fig. 1.1b	in logarithmic transformation
Fig. 2.1	by classical methods
Fig. 4.6	by high-order polynomial
Fig. 5.2	by first-order polynomial
Fig. 5.8	by second-order polynomial
Fig. 5.12	by third-order polynomial
Fig. 6.10	by Richards function
Fig. 7.1	by segmented polynomials
Fig. 7.3	by running re-fit
Figs 7.4 and 7.5	by splined regression

In addition, we have seen classically determined progressions of leaf area ratio and unit leaf rate (Fig. 2.3), comparisons involving the growth of this variety in four different years (Fig. 7.6) and will shortly see two further and more special presentations (Figs 8.1 and 8.3). If we follow through the arguments variously presented at the first appearance of each of these analyses, we find that the only truly satisfactory one has unquestionably been the last and most complex of these, the splined cubic polynomial exponential. This brings us to another question.

8.2 The classical versus the functional approach

> 'The advantages of curve fitting as presented . . . suggest that the classical approach to growth analysis might well be buried in the sands of time.'
> F. I. Woodward (1980)

8.2.1 Contest or not?

The above was written in a review of *Plant Growth Analysis* (Hunt, 1978a) which, naturally, had given due coverage of the functional approach. It highlighted an important question to which everyone who grows plants under the circumstances mentioned in the Introduction (p. 2) will require an answer before his experimental design can be completed. Woodward himself was equivocal on the issue, for the quotation continues

> 'However, . . . comparisons of classical and curve-fitting analyses of the data of Kreusler *et al*. . . . suggest that the curve fitting techniques require to be complex before they can be the equal of calculations by classical techniques, particularly over long experimental periods.'

The Kreusler series listed in section 8.1.2 appears to confirm Woodward's caveat even more strongly. Is this true; and, if so, is it general?

8.2.2 The relevant issues and the Kreusler data again

In section 3.3 we saw, in addition to what was claimed as the one supreme advantage of the functional approach to plant growth analysis, twelve subsidiary advantages. These are each real enough, and any one might alone be a decisive feature according to the needs of the experimenter. So any one, 'more complex' or not, might lead to an improved analysis in any individual case. However, on a great many occasions it may be found that the most beneficial advantage to be gained from the functional as opposed to the classical approach will involve the issue of smoothing. Connected with this we saw in our list of advantages: (4) 'assumptions evaded', (5) 'all data involved', (6) 'pairing unnecessary' and, of course, (10) 'smoothing possible'.

The enhancement that smoothing brings, or even the requirement which arises that it be done, springs both from the nature of the primary data themselves and from the way in which these are collected. Here we remember that if it could be followed exactly, the diurnal course of weight increment in green plants would exhibit sizeable exogenous fluctuations, reflecting the gross courses of carbon fixation and the uptake of water and nutrients. But because the taking of dry weight involves destroying the plant, such studies as already exist in this field have largely been confined to non-destructive observations on change in fresh weight or in linear dimensions (see Hunt, 1980, for references). Change in dry weight, though of course more useful, has conveniently had to be studied over periods of days or weeks by means of a series of destructive harvests spaced at multiples of whole days, thus negating diurnal fluctuations by positioning each sample at the same point in the diurnal cycle of plant activity. This has been necessitated by the serious practical difficulties that stand in the way of more frequent sampling for dry weight and by the fact that such data, when obtained from closely spaced harvests drawn from natural populations of experimental plants, tend to be statistically 'noisy' (in the sense that even the sign of dry weight change between adjacent harvests is often in doubt).

But, even when sufficient growth has been made in the interval between two harvests for the classical approach to be a practical proposition (Causton and Venus, 1981, p. 37), another problem still remains. With this in mind, Hunt and Evans (1980) drew attention to

> 'a feature of the conventional method of calculating mean values of unit leaf rate over a period (Williams, 1946) which involves subtracting the dry weight at the beginning from that at the end: that of a seesaw of successive values in a time sequence. For if the dry weight at the end of a particular period is by chance high, then the calculated value of unit leaf rate for this period will also be high. But the same high dry weight applies to the beginning of the next period and will by the same token diminish the estimate of unit leaf rate in that period, and vice-versa. . . . This feature complicates the study of individual values of mean unit leaf rate and their general trends, and a comparison of Briggs, Kidd and West's Figure 3 [1920] with our Figure 7 (which uses the same primary data) shows how much clearer is the presentation when instantaneous values are used, taking account statistically of the characteristics of the progressions of the primary data over longer periods.'

Hunt and Evans's 'Figure 7' is reproduced, in part, as the present Fig. 8.1b. This contains instantaneous unit leaf rates, **E** (from the cubic spline analysis described in section 7.4.1), here plotted against the background of the harvest-interval means, taken from Fig. 2.3; Fig. 8.1a shows the same for leaf area ratio, **F**. All are for Kreusler's maize grown in 1878 (Table 1.1). The fitted progression of **F** corresponds to that of $\overline{\mathbf{F}}$ as well as could ever be hoped for, each value of **F** lying more or less mid-way between the adjacent levels of $\overline{\mathbf{F}}$. In **E**, however, the fitted progression steers a parsimonious course among values of $\overline{\mathbf{E}}$ which amply illustrate the seesawing highlighted by Hunt and Evans (1980). And we remember that the variates actually fitted in this analysis were $\log_e W$ and $\log_e L_A$, not $\overline{\mathbf{F}}$ or $\overline{\mathbf{E}}$. In one sense we are thus fitting 'at a distance' but, in another, we are reflecting the true relationships between the primary data taken as a whole and not their erratic coincidences about individual harvest-intervals.

In the case of **F**, then, the functional approach has not enhanced our perception of the true course of this derivate beyond that which can be supplied by classical methods, even though there may be a separate, and even decisive, statistical advantage in treating the whole progression as one. This lack of difference is due partly to the nature of the derivate, $\overline{\mathbf{F}}$, itself

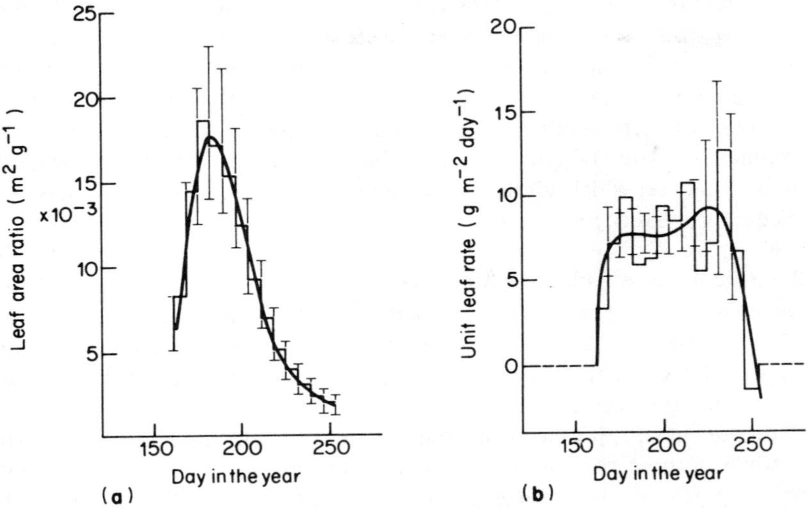

Fig. 8.1 Progressions of **(a)** leaf area ratio and **(b)** unit leaf rate, both with 95% limits and harvest interval means, derived from splined cubic polynomial exponentials fitted to Kreusler's data for maize (Table 1.1), see also Fig. 7.5 (analyses from Hunt and Evans, 1980).

and partly to its method of calculation, as Hunt and Evans (1980) explain. But in the case of **E**, there can be no doubt that the instantaneous progression is greatly enhanced representation of the underlying trend, of which the values of $\overline{\mathbf{E}}$ are but a poor shadow of reality (for reasons of chance unrepresentativeness or genuine short-term variation – we cannot really say just how much of which, though Hunt and Evans (1980) looked for environmental correlations). Not only this, but there can also be no question that such advanced methods of analysis are unnecessary to achieve this goal, all others, as we have seen (and this is even more true for **E** than for **R**), falling short in some major way of a satisfactory representation of trends within this most testing of data sets.

So, returning to, but not yet finishing with, Woodward's comment, we see that, at least in respect of **E** (and, of course, of **R**), the functional analysis (the splined cubic polynomial exponential) is not the equal but the superior of the classical. In less complex data sets, less complex analyses would achieve the same net gain.

But, what if less complex functional analyses are applied to complex primary data? Is there some point in a continuum of functional complexity at which an equality of 'value' is achieved between the functional and the classical approaches, and below which the classical can be considered superior? No, not always, for the two approaches are also sufficiently unlike in the information that they supply for there often to be no possibility of parallel comparisons on this basis, even in data sets which admit to both, and not all do that. Rather, the situation is normally that, even given identical objectives, the two methodologies are different in kind, not in degree. Each possesses its own set of consequences which must be balanced the one against the other by the experimenter. And, of course, should the experimenter be forced into, or inherit, a data set which admits to only one of these two approaches, then his choice is removed anyway.

8.2.3 Comparisons within the literature

In addition to the many theoretical papers cited in Chapter 3, and by Hunt (1979), there have been several which have included a worked comparison of the classical and functional approaches as a major feature of their presentation. These are listed in Table 8.1.

None of these, however, is illustrated here. Instead, Fig. 8.2 is given with the triple purpose of saluting one of the very earliest comparisons of this type, supporting one of the examples of a logistic fit (Fig. 6.3), and illustrating the high desirability of employing the functional approach where extensive data sets are involved (section 8.2.3).

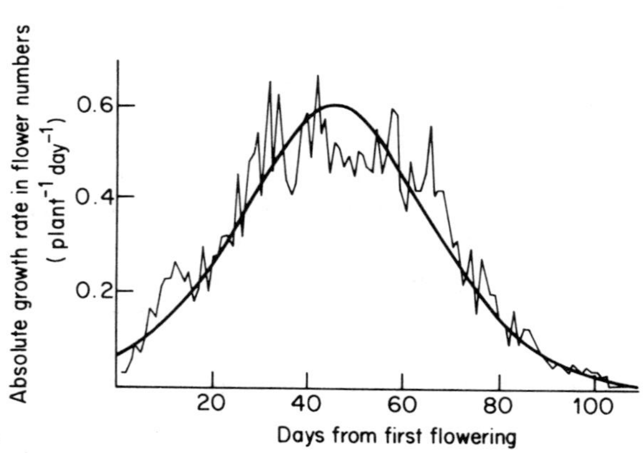

Fig. 8.2 Classical and functional analyses of the flowering curve in cotton, var. Sakellaridis, sown at Bahtim (Egypt) in March 1920 (see Fig. 6.3). The smooth curve is the absolute growth rate in flower numbers per plant derived from a fitted logistic; the zig-zag line joins the daily mean values of this derivate (data and analyses from Prescott, 1921).

8.2.4 Too few data for the functional approach

Having done all we can for the time being with Kreusler's data (Table 1.1) we bid them farewell in one final example of how, were they not so numerous, the situation, both analytically and biologically, would be very different.

Table 8.1 Studies involving substantial comparisons of the classical and the functional approaches to plant growth analysis.

Author(s)	Species	Time interval	Primary data	Derived	Comments
Hammerton and Stone (1966)	*Polygonum lapathifolium* (pale persicaria)	22–92 days (p)	Ws, L_A, N	**Rs, F, E, A**	Various functions also compared
Buttery (1969)	*Glycine max* (soybean)	30–170 days (p)	W, L_A	**Rs, E, C, L**	Full statistics given
Eckardt *et al.* (1971)	*Helianthus annuus* (sunflower)	43–106 days (p)	W	**C**	Calorimetry and gas analysis also compared
Silsbury (1971)	*Lolium perenne* (perennial ryegrass)	32 days (p)	W, L_A	**Rs, E**	
Buttery and Buzzell (1974)	*Glycine max* (soybean)	21–84 days (p)	W, L_A	**E**	
Hunt and Parsons (1977)	*Zea mays* (maize)	113 days (g)	W	**R**	See Fig. 7.1
Hurd (1977)	*Lycopersicon esculentum* (tomato)	10–48 days (p)	W, L_A	**Rs, F, E**	See section 5.6
Hunt and Evans (1980)	*Zea mays* (maize)	Up to 108 days (e)	W, L_A	**Rs, F, E**	See Fig. 8.1
Causton and Venus (1981)	4 spp.	Various	Ws, L_A	**Rs**, ratios, **E**	
Parsons and Hunt (1981)	*Zea mays* (maize)	113 days (g)	W	**R**	See Fig. 7.5

For footnotes see Table 5.1.

Figure 8.3 shows analyses not of the whole data set but of edited selections of them. For these, the policy was for part (a), take one, miss seven; and for part (c), take one, miss three. Parts (b) and (d) are the corresponding progressions of **R** and the classically derived $\overline{\mathbf{R}}$. Analyses were by the stepwise methods of Hunt and Parsons (1974), see section 5.6.

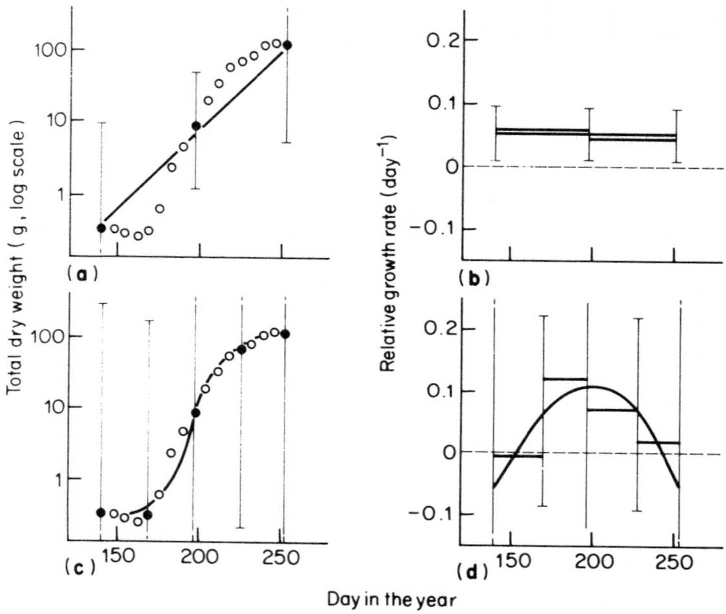

Fig. 8.3 Kreusler's data for maize (Table 1.1) edited in two special ways before analysis: **(a)** to leave three data only (solid symbols), eliminating the intervening series of seven data (open symbols), **(c)** to leave five data only, eliminating the intervening series of three data; the surviving data in **(a)** are fitted by a first-order polynomial exponential, with 95% limits and those in **(c)** are fitted by a third-order polynomial exponential, again with 95% limits (mostly off-scale); parts **(b)** and **(d)** depict progressions of instantaneous and mean relative growth rate, the former with 95% limits, derived respectively from **(a)** and **(c)**.

Figure 8.3 shows the wealth of detail that is missing from such a progression and underlines the necessity for having observations of sufficient frequency to match the expected level of detail in the process under study. The plateau in **R** which results from the first-order polynomial exponential matches the values of $\overline{\mathbf{R}}$ well but, of course, it recognizes none of the slight downward ontogenetic drift present in $\overline{\mathbf{R}}$, which itself only hints at the 'true' situation, as represented by the complete data set. In Fig. 8.3c we have a fragile, but recognizable, sketch of the underlying progression in $\log_e W$, but that of **R** fails to match the left-skewed progression of $\bar{\mathbf{R}}$, so again the latter remains the more appropriate analysis (an analysis conducted on a policy of take one, miss one was found to be virtually identical). Figures 8.3b and 8.3d both illustrate the large errors of derivates that commonly occur during over-fitting. So, here we have artificially created instances of under-represented progressions; what limited perception of these is possible, is only so by way of classical methods.

Then there is the situation where there is not lack of fit, as in Fig. 8.3c, but too much, due either to there being too few data or too powerful a statistical instrument used to fit them. The disadvantages of the perfect fit with no smoothing at all have already been discussed (sections 4.6, 5.5) but there is one even greater danger that occasionally arises, that of the perfect fit by ridiculous means. Pollard (1977) exemplified this deliberately and neatly with a splined regression fit in the form $X = f(\log_e X)$. Figure 8.4 shows how this function traverses every datum, and shows also that that is just about all that can be said for it. This analysis would clearly be inferior to the classical approach and inferior, too, to saner functional methods.

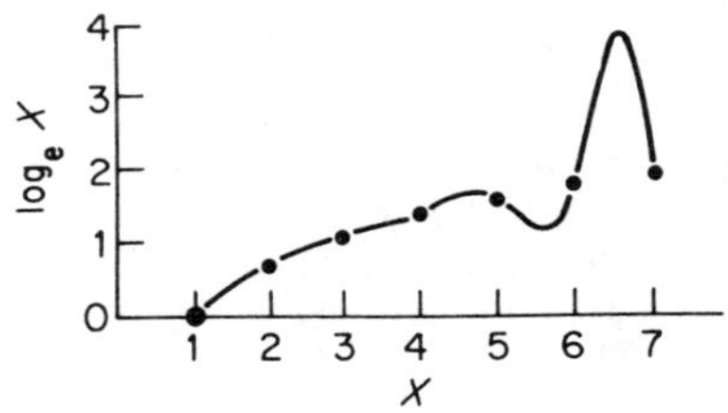

Fig. 8.4 An absurdly over-fitted progression: quadratic splines fitted to seven values of $\log_e X$ plotted against X (given as a cautionary analysis by Pollard, 1977).

8.2.5 Too many data for the classical approach

In 1978 Monselise, Varga and Bruinsma published a remarkable series of observations on the growth in volume of the individual fruits of *Lycopersicon esculentum* (tomato). Using data originally collected by Varga and Bruinsma (1976), data involving daily measurements on up to 35 fruits for 54 days, they presented the progression of mean total volume per fruit and solved equations 2.2 and 2.7 more than fifty times apiece to obtain day-to-day values of $\bar{\mathbf{G}}$ and $\bar{\mathbf{R}}$ respectively, progressions of which they then displayed (Fig. 8.5). Moreover, they then plotted these values of $\bar{\mathbf{R}}$ on a logarithmic scale and showed by means of the resulting linearly declining slope, that the relative decay rate of relative growth rate (more than the second derivative of $\log_e W$, see section 3.4.7), was effectively constant after the first week of growth. While applauding this advanced and informative analysis, it cannot be

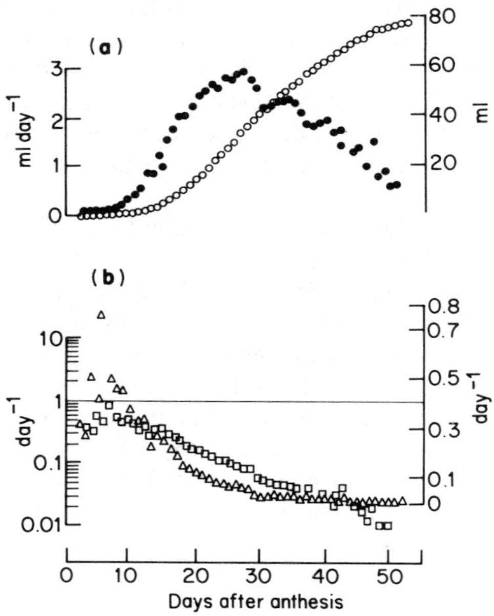

Fig. 8.5 Progress curves depicting the growth in volume of the tomato fruit: **(a)** cumulative volume per fruit (○); absolute growth rate in volume per fruit (●); **(b)**, relative growth rate in volume per fruit (△); relative growth rate in volume per fruit, on a logarithmic scale (□). All derivates are classically determined (data from Varga and Bruinsma, 1976; analyses from Monselise, Varga and Bruinsma, 1978).

denied that the use of fitted curves, possibly the Gompertz, Richards or cubic splines, would have added greatly to the stability of the derivatives obtained and also permitted rigorous statistical analysis. This was thus an instance not of an invalid, but of an unnecessary and suboptimal application of the classical methodology.

Other lengthy series of data which would have benefited, in particular, from splined curve fitting have been presented by Cannell and Willett (1976), Gerakis and Papakosta-Tasopoulou (1979) and Hurd, Gay and Mountifield (1979), the last of whom studiously avoided curve fitting of any type despite their substantial previous activity in this field. In need of asymptotic fits were the data sets of Montenegro, Aljaro and Kummerow (1979), Mukai *et al.* (1979), Simmons and Crookston (1979), and Wardlaw, Sofield and Cartwright (1980), while in need of response surface analysis in three dimensions (section 7.7) was the data matrix of Solárová (1980) in which replicated measurements of epidermal diffusive conductance in leaves of *Phaseolus vulgaris* (bean) spanned sixteen observational occasions at each of eight levels of irradiance.

8.3 The biological relevance of parameters and functions

> '. . . mathematical functions based on simple hypotheses concerning the nature of growth are capable of reproducing the course of growth curves with tolerable accuracy, but not so accurately that a clear distinction can be drawn between one function and another. A sounder approach is to look for accuracy of fit, and to reject the notion that the mathematical form has physiological significance.' R. F. Williams (1975) (after F. J. Richards, 1969)

It may well be that we have already seen enough finger-pointing from these pages to allow the very large number of authors who have chosen, either openly or tacitly, to ignore Williams's (and Richards's) advice to go unlisted. The usual viewpoint advanced to support this reticence, either for asymptotic functions instead of polynomials or for no fitting at all instead of some, is that parameters or whole functions that cannot be ascribed any direct biological significance cannot be of any biological value.

Doubtless this viewpoint has been inherited from the physical sciences where components having clear mathematical analogues fit together, either figuratively or literally, like cogs in a machine (section 3.2). And if this viewpoint's passage into plant growth studies has been by way of animal science (see Zuckermann, 1950) then that experience, too, will have coloured its adherents' expectations. In our own field, these expectations are falsely encouraged by the fact that some of the asymptotic functions, such as the logistic, and the simpler of the polynomials (the first- or second-order) have parameters to which can be ascribed some rather general type of biological significance in terms of rate of increase, acceleration, initial or final size, and so on. However, not even these have any direct and identifiable analogue in the plant world, their only virtue being that they may themselves be used as comparative tools which apostrophize processes and properties which, while not being entities in themselves, have some general and collective interest. Parameter values in higher-order polynomials, and in most of the approaches described in Chapter 7, have no such virtues. Therefore, runs the argument, they are useless.

But, to reject the whole approach on these grounds alone is, to put it charitably, unwise. For the comparative information is supplied not through the functions' parameters but through their derivates. For example, nowhere in our preferred analysis of Kreusler's data have we found it necessary or even desirable to refer to parameter values in the splined cubic polynomial exponentials, despite the fact that it is these alone which transform the primary data of Table 1.1 into the enhanced representations given in Figs 7.5 and 8.1. Such parameters are messengers of reality, not reality itself; to expect otherwise would be both naïve and lazy. After all, no-one else eschews radio telecommunications because the signals themselves are, on first acquaintance, found to be incomprehensible.

So, whether or not the empirical function fitted to observational data, in the terms of Thornley's (1976) description, 'tells one nothing that is not contained in the data' is open to question. Looking at Fig. 8.1a we are inclined towards one conclusion on this issue: looking at Fig. 7.7a we are inclined towards quite another.

8.4 Why plant growth analysis?

> '. . . properly handled, growth analysis is at best a useful method for obtaining preliminary descriptions of plant growth as a framework for more detailed studies of the underlying biological processes, and at worst an end in itself leading on to nothing.' H. W. Woolhouse (1980)

8.4.1 Two awkward questions

At the beginning of this work I suggested (in section 1.4) that plant growth analysis occupied a clearly recognizable niche within the extensive spectrum of activity that now involves the quantitative study of plant growth. Surveying this scene from the ecosystem level of organization to that of the organ, and beyond, we saw that the traditional role of plant growth analysis had been played at the level of the population and of the whole organism and that, primarily, this role involved the derivation of indices of dry matter production, leafiness and leaf efficiency. From these observations, and prompted by Woolhouse's stern assertion (an extract, in fact, from another review of Hunt (1978a) *Plant Growth Analysis*), two major questions now arise. First: What is the justification for continuing at these levels with the traditional technique of plant growth analysis, either classical or functional, when many of its original objectives have been superseded by more recent and elaborate methods? Second: Is there any new, or more unorthodox, role to be played by plant growth analysis, either at these levels of organization or beyond them?

8.4.2 Dodo or doyen?

There is no doubt as to the high seniority of plant growth analysis among quantitative methods of investigating plant performance. The work of Gregory, Blackman and Briggs in the 'teens of this century, with that of their antecedents all ably chronicled by Evans (1972, p. 190 ff.), shed floods of new light upon our appreciation of how fast plants work and upon our knowledge of plasticity in their form and function at various stages throughout their life cycles. One of the principal new advantages gained as a result of this work was a valuable long-term measure of apparent assimilation, the unit leaf rate. This exemplifies well the mild controversy that has lately built up around plant growth analysis, since it is this measure which has suffered the most neglect, and even animosity, at the hands of many plant physiologists as, little by little, they have been able to arm themselves with the means of determining more directly the plant's net balance in fluxes of carbon dioxide, oxygen or water vapour, hence achieving laboratory-based estimates of photosynthetic production at a more fundamental level.

Now, this trend has been a noteworthy manifestation of reductionist theory in plant science: the notion that, mechanistically speaking, if one looks after the pennies the pounds will look after themselves: otherwise, understand molecules and you will understand ecosystems. While the truth in this might be evident to a visiting Martian, the more astute observers of the world of plants will be inclined to agree with A. P. Hughes, who once remarked to the author, in a different, but not unrelated, context that

> 'It is senseless to attempt to empty the Channel with a teacup however desirable a land crossing to France may be.'

Naturally, if the objective is to study the behaviour of seawater in teacups then this methodology is appropriate; normally, however, some grander aim is professed.

Increasingly, and not really surprisingly, this reductionism is now being questioned. For example, in a perceptive essay on research strategy in plant science, by Thornley (1980), we can read that

> 'Observation of the research scene in biology leads one to conclude that most biological scientists are reductionists by nature. People feel that, if only they could understand the level below the one at which they are working, then all their problems would be solved. This has resulted in a situation where often agronomists tend to do crop and plant physiology; in some departments of plant physiology one finds people doing biochemistry or even molecular biology; and likewise in departments of biochemistry and molecular biology, work in physics or physical chemistry may be in progress. Of course this overstates the case, and no doubt there are some benefits accruing from the present arrangements. However, it is possible that a blinkered approach, of mostly looking downwards, may be leading to a use of research resources that is less effective than it might be.'

Then, Passioura (1979), in a powerful analogy concerning the imaginary scientific investigation of a photograph printed in a newspaper, has warned of how dangerous such reductionism could be. He feared that in such an investigation the modern scientist would be drawn into a detailed description of the myriad of dots that make up the printed photograph, analysing number, size, distribution and pattern, and neglecting entirely the properties of the system at higher levels of organization: light and shade, balance and perspective. He could have added, too, that at higher levels still, social or political information might also be found. All done by dots, yes, but not by dots alone. Again, Hunt (1981), in discussing the balance between mechanism and empiricism in the modelling of plant growth, also saw this balance as hinging principally upon level of organization. Empiricism and mechanism were held to be two sides of the same coin:

> 'For example, the organismal biologist will regard the work of the population biologist as empiricism since it takes no account of mechanism at the level of organization in which he is interested. Similarly, his own work, which he may well describe as mechanistic, may be regarded as empiricism by the cellular biologist who seeks reality at a still lower level of organization. We have a Janus whose aspect depends upon the standpoint of the observer. In absolute terms there is no good or bad in either approach, the sole criterion is suitability for the purpose in hand'

and proper awareness of, and sympathy with, work at other levels of organization.

Where, then, does all of this leave the traditional corpus of plant growth analysis? I see its position here as being somewhat akin to that of railway transport: just because some of its original functions have been superseded by more recent developments, this does not mean that none remains, or that any which do are any less special, or incapable (or unworthy) of further evolution.

So, when assessing which facets of plant growth analysis have been overtaken and which have remained useful, we must firstly take note of the fundamental nature of growth-analytical quantities: they describe the situation *de facto*, integrated both throughout the plant and across substantial intervals of time. In short-term studies in environmental physiology, for example of plant photosynthetic production, the plant is, as I have suggested in section 1.4, judged less by results than by promises. Thus, while it is certainly true that, for example, unit leaf rate can now be bettered by derivations from gasometric observations when we seek short-term measures of apparent assimilation (inasmuch as unit leaf rate was ever intended to be this), it cannot at present be replaced by any equally simple measure of long-term photosynthetic efficiency. Evans (1976), in dealing with the time element as a reason for discounting so much of this type of short-term observation, warned of

> 'the series of problems posed by the necessity of integrating observations made over periods of the orders of minutes and hours so that they can take their place in a picture of the life of the plant whose growth is measured in terms of weeks, months or years. These problems are of great complexity, and attempts to solve them without reference to the growth of the whole plants themselves in nature have rarely been successful. Why not then allow the plant to do its own integration undisturbed by experimental interference? The experimenter can easily make mistakes in assessing how the plant at a given stage of development would respond to a given change in the external environment. The undisturbed plant growing in its normal natural surroundings is unlikely to do so.'

Then, in addition to the time element, there are questions in environmental physiology which concern representativeness, interference and trauma – questions vital to the validity of ecological or agronomical experimentation. It is inherently in the nature of the method that these are much more easily settled within the context of plant growth analysis: here, the growing plant need be disturbed by the experimenter only once, and when that call comes its work is done. Even in controlled environments, uniformity and repeatability are not won at the cost of disturbance. However, in environmental physiology, whatever liberties may be taken with the plant in purely physiological work, it is not uncommon to read in an *ecological* journal of part of a plant being excised from its parent, transported great distances in the dark, re-cut under distilled water, stored in a controlled environment, shorn of further components in the interests of standardization and practical convenience and then positioned in an artificial assimilation chamber in the hope of recording an ecologically meaningful performance. That such a performance may easily be observed and repeated adds nothing to its validity: circus tricks are that, too. As with certain types of documentary television, if the technique of gaining access destroys or modifies the subject, what do we learn? Craft, skill and understanding are needed to protect reality from its investigator.

Evans (1976) recognized this, too, and more besides. To him,

> 'It was early apparent that . . . the study of undisturbed growing plants had the unique advantage of being the only system of study which could at need deal with the natural plant/environment system without external interference over periods of the order of a week. Thus it not only had great advantages from the point of view of integration over long periods, but also offered the possibility of constructing a quantitative framework of reference into which other quantitative studies of parts of the whole system could be fitted with confidence.'

But, having drawn this considerable distinction between the nature of growth-analytical and environmental-physiological study, it may also be said that it is now high time that the two were overlapped to the point of fusion. Hunt (1980) showed that it is no longer true that 'a minimum period of several days' (Šesták, Jarvis and Čatský, 1971) is necessary for the effective employment of growth-analytical methods, and for example, Šetlík and Šesták (1971) and Hadley *et al.* (1980) have explored the possibilities of long-term gasometric measurements of photosynthesis spanning several days. Practical difficulties, and even politico-scientific differences, have succeeded in keeping apart activities which are long overdue for coalescence.

To generalize brutally: even considering the organismal level alone, plant scientists have generated a wide spectrum of activity; at the growth-analytical end there is valid work lacking in detail and at the environmental-physiological end, detailed work lacking in validity. Which is truer science we will not ask. In some areas we have evidence in abundance of work done at the two ends of this spectrum. To render it seamless what we now require is more work from its centre: more short-term growth-analytical studies, protected by unusually stringent precautions against artefact, and more long-term and wholly field-based environmental physiology, gaining ecological validity at the cost of great inconvenience. For both, the days of their easy pickings may in some senses be over.

The question posed in the title of this section is thus answered: plant growth analysis, though old, is still useful. Many of its original capabilities remain uniquely valuable and, properly handled and interpreted, form an indispensible adjunct to more lately arrived methodology, the undisputed gains of which have exacted their price in ecological verity.

8.4.3 A wider future?

We saw in section 2.1 that all of plant growth analysis could be resolved into matters involving one of four general types of quantity,

(i) simple rates of change
(ii) simple ratios between two quantities
(iii) compounded rates of change
(iv) integral durations,

and that one frequent type of inter-relationship between these was seen to be (i) = (ii) × (iii). To me, the future of plant growth analysis rests not only upon its advantageous treatment of the problems posed by time-scale and by human interference, but also upon the inextinguishable utility of relationships of the foregoing type. To use the terminology of economics, the notions that industrial production springs jointly from staffing level and productivity (per man) or, on a national scale, that gross national product depends jointly upon the level of employment and upon that same productivity, are of great intellectual accessibility. In plant growth analysis, the anthropocentric view that results are achieved by a workforce of a certain size functioning at a certain rate per individual, has long provided a valuable point of entry into the mysterious and inaccessible world of plant function. This advantage has already been demonstrated at many levels of organization and to this I see no sign of an end. On the contrary, when it is remembered that in the landscape of biology each entity is composed of a number of subentities all functioning in their own particular ways to contribute to a greater whole, we see that obscurity still reigns amidst our knowledge of the quantitative links between the entities we see displayed hierarchically within this vista. Rich applications of the concepts of production, functional size and efficiency still await exploitation. Physical sciences, though in many ways far more amenable to our type of approach, have not the advantage of an hierachical organization of this extent.

In addition, the individual quantities involved in plant growth analysis naturally possess their own merits. For example, we have already seen nine valuable guises in which quantities of type (iii) may appear (sections 2.2.4 and 2.2.7). The biological entities that produce other biological entities do so at certain rates, so these 'rates of production of something per unit of something else' become tools of universal and substantial importance in plant science, quite irrespective of level of organization. Then there is the opportunity of relating these concepts to other fields of study, not normally associated with plant growth analysis. One such association may be described in order to illustrate the advantages of this type of cross-fertilization.

Grime (1974, 1977, 1979) has advanced a comprehensive interpretation of the processes which, in the established phase (that is, separately from the question of regenerative ability), control the structure and composition of natural vegetation. In essence, Grime's thesis implies this: the inherent procreative tendency of plant life is opposed by an environmental force which is hostile to the production of plant biomass. On the outcome of each particular conflict between these opposing trends depends the detailed structure and composition of vegetation.

To gain definition, the hostile force is resolved into two vectors, stress and disturbance; either may be biotic or abiotic in origin, or both. Stress, Grime determines as

> 'the external constraints which limit the rate of dry matter production of all or part of the vegetation'

and disturbance as

> 'the mechanisms which limit . . . plant biomass by causing its partial or total destruction'

one operating broadly prior to, and one broadly after, plant growth. These two may vary both in quality and in strength and, within any one vector, each of these two variables (quality and strength) may exert its influence either alone, or in combination with the other. Irrespective of quality, if either stress or disturbance is strong, or if neither is, then vegetation of some kind or other can result; if both vectors are strong then the habitat is impossible for plant life. In purely quantitative terms, the combination of low stress/low disturbance is held to favour plants of a 'competitive' evolutionary strategy; that of low stress/high disturbance, plants of a 'ruderal' strategy; and that of high stress/low disturbance, plants of a 'stress-tolerant' strategy. Each particular combination leads to one of a diverse family of vegetation types, according to the quality of the vectors currently in operation, but all members of this family remain linked by the same recognizable traits, which are governed by the vectors' quantitative strength. Further, the quality of either the stress or the disturbance may itself lead to a characteristic vegetation type, if its influence is sufficiently strong.

The three strategies represent the poles of a triangular system of ordinating plant species. This is seen as a pure continuum within which several intermediate or 'secondary' strategies can be recognized. Thus constructed, Grime's interpretation spans both woody and herbaceous communities, and also systems containing non-vascular species. Where plant growth analysis comes in is that relative growth rate in the seedling phase of the life-cycle (the $\mathbf{R}_{max}$ of Grime and Hunt, 1975, see section 2.2.2) has proved to be the best general indicator yet encountered of plant adaptation to the stress vector in

Grime's scheme. The relationships involved here are that high potential relative growth rate is positively advantageous in conditions of low stress for plants of the 'competitive' and 'ruderal' strategies (the difference between these two being in their response to disturbance) and positively deleterious in conditions of high stress, where plants of the 'stress-tolerant' strategy may find a refugium from the harmful competition of faster-growing neighbours because the former's inherently lower relative growth rates do not deplete the limited resources of their environment so rapidly as would plants of either of the other two strategies. Hence, stress-tolerant plants sustain a fragile existence in situations that lead to disaster for more voracious species.

However, Grime and Hunt's (1975) $\mathbf{R}_{max}$ may still be only a rough-and-ready approximation to this notion of an ecologically-adaptive 'growth rate', especially when it is remembered that the true course of growth in dry weight in seedlings of an herbaceous native species, raised more or less unrestrictedly in a productive environment, looks something like that given in Fig. 8.6.

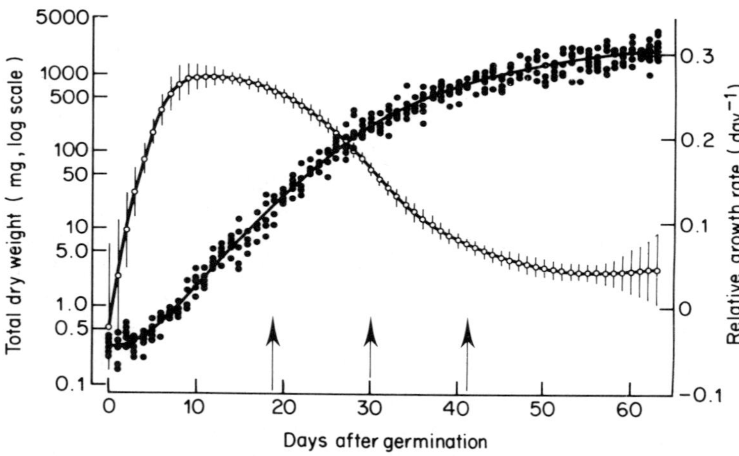

Fig. 8.6 An extended series of data on the growth in total dry weight of spaced individuals of *Holcus lanatus* in a productive, controlled environment. Closed symbols represent the primary data (one per plant) which are fitted by a three-knot spline using the methods of Parsons and Hunt (1981); knot positions are indicated by arrows; open symbols are relative growth rates derived, with 95% limits, from the fitted spline; see also Figs 7.7, 7.8d (adapted and extended from data and analyses given by Hunt, 1980).

Where in this is our parameter to be found? In the acceleration to maximum **R**? In the value of the maximum **R**? In the earliness of the maximum **R**? In the time over which **R** is at least fifty per cent of $\mathbf{R}_{max}$? In $\overline{\mathbf{R}}$ over a certain period? In the time over which **R** is significantly above zero? All of these questions deserve an answer.

And last of all, lying behind all of the hierarchical activity previously described, and behind the multitudinous ramifications which can be developed at any individual level, such as the above case involving evolutionary strategies, there is the inescapable fact that, at nearly all levels, plant growth is a dynamic, largely continuous and mathematically form-free process which, for the many reasons given in section 3.3 (and by Hunt, 1979), and as our final figure amply demonstrates, gains enormously in definition when the reality of the underlying time-course of the process under study is recovered through the agency of the fitted plant growth curve.

Literature cited

ABUL-FATIH, H. A., BAZZAZ, F. A. and HUNT, R. (1979). The biology of *Ambrosia trifida* L. III. Growth and biomass allocation. *New Phytologist,* **83**, 829–38.

ACEVEDO, E., FERERES, E., HSIAO, T. C. and HENDERSON, D. W. (1979). Diurnal growth trends, water potential, and osmotic adjustment of maize and sorghum leaves in the field. *Plant Physiology,* **64**, 476–80.

ADEPETU, J. A. and AKAPA, L. K. (1977). Root growth and nutrient uptake characteristics of some cowpea varieties. *Agronomy Journal,* **69**, 940–3.

AFZAL, M. and IYER, S. S. (1934). A statistical study of the growth of main stem in cotton. *Indian Journal of Agricultural Science,* **4**, 147–65.

AHLBERG, J. H., NILSON, E. N. and WALSH, J. L. (1967). *The Theory of Splines and their Applications.* Academic Press, New York.

AHLOOWALIA, B. S. (1973). Germination in vitro of reygrass pollen grains. *Euphytica,* **22**, 575–81.

ALEXANDER, R. McN. (1971). *Size and Shape.* Studies in Biology No. 29. Edward Arnold, London.

ALLEN, R.L. (1976). Method for comparing fish growth curves. *New Zealand Journal of Marine and Freshwater Research,* **10**, 687–92.

ALLISON, J. C. S. (1969). Effect of plant population on the production and distribution of dry matter in maize. *Annals of Applied Biology,* **63**, 135–44.

ALLISON, J. C. S. (1971). Analysis of growth and yield of inbred and cross-bred maize. *Annals of Applied Biology,* **68**, 81–92.

ALLISON, J. C. S. and WATSON, D. J. (1966). The production and distribution of dry matter in maize after flowering. *Annals of Botany,* **30**, 365–81.

AMER, F. A. and WILLIAMS, W. T. (1957). Leaf-area growth in *Pelargonium zonale. Annals of Botany,* **21**, 339–42.

ANALYTIS, S. (1974). Der Einsatz von Wachstumsfunktionen zur Analyse der Befallskurven von Pflanzenkrankheiten. *Phytopathologische Zeitschrift,* **81**, 133–44.

ANDERSON, O. D. (1976). *Time Series Analysis and Forecasting: the Box-Jenkins Approach.* Butterworth, London.

ANDERSON, O. D. (1977). A commentary on 'A Survey of Time Series'. *International Statistical Review,* **45**, 273–97.

ANGUS, J. F., NIX, H. A., RUSSELL, J. S. and KRUIZINGA, J. E. (1980). Water use, growth and yield of wheat in a sub-tropical environment. *Australian Journal of Agricultural Research,* **31**, 873–86.

ARNOTT, R. A. (1975). A quantitative analysis of the endosperm-dependent seedling growth in grasses. *Annals of Botany,* **39**, 757–65.

ASHBY, E. (1929). The interaction of factors in the growth of *Lemna.* III. The interrelationship of duration and intensity of light. *Annals of Botony,* **43**, 333–54.

ASHBY, E. (1930). Studies in the inheritance of physiological characters. I. A physiological investigation of the nature of hybrid vigour in maize. *Annals of Botany,* **44**, 457–67.

ASHBY, E, (1932). Studies in the inheritance of physiological characters. II. Further experiments upon the basis of hybrid vigour and upon the inheritance of efficiency index and respiration rate in maize. *Annals of Botany,* **46**, 1007–32.

ASHBY, E. (1937). Studies in the inheritance of physiological characters. III. Hybrid vigour in the tomato. Part 1. Manifestations of hybrid vigour from germination to the onset of flowering. *Annals of Botany,* **1**, 12–41.

ASHBY, E. and WANGERMANN, E. (1950). Studies in the morphogenesis of leaves of *Ipomoea* in relation to their position on the shoot. *New Phytologist,* **49**, 23–35.

ASIMI, S., GIANINAZZI-PEARSON, V. and GIANINAZZI, S. (1980). Influence of increasing soil phosphorus levels on interactions between vesicular-arbuscular mycorrhizae and *Rhizobium* in soybeans. *Canadian Journal of Botany,* **58**, 2200–5.

ATTIWILL, P. M. (1979). Nutrient cycling in a *Eucalyptus obliqua* (L'Hérit.) forest. III. Growth, biomass and net primary production. *Australian Journal of Botany,* **27**, 439–58.

AUSTIN, R. B. (1963). A study of the growth and yield of carrots in a long-term manurial experiment. *Journal of Horticultural Science,* **38**, 264–76.

AUSTIN, R. B. (1964). A study of the growth and yield of red-beet from a long-term manurial experiment. *Annals of Botany,* **28**, 638–46.

AUSTIN, R. B., NELDER, J. A. and BERRY, G. (1964). The use of a mathematical model for the analysis of manurial and weather effects on the growth of carrots. *Annals of Botany,* **28**, 153–62.

BAILEY, N. T. J. (1964). *Statistical Methods in Biology.* English Universities Press, London.

BAKER, C. H., HORROCKS, R. D. and GOERING, C. E. (1975). Use of the Gompertz function for predicting corn leaf area. *Transactions of the American Society of Agricultural Engineers,* **18**, 323–30.

BAKER, E. F. I. (1979). Mixed cropping in Northern Nigeria. III. Mixtures of cereals. *Experimental Agriculture,* **15**, 41–8.

BARBER, S. A. (1978). Growth and nutrient uptake of soybean roots under field conditions. *Agronomy Journal,* **70**, 457–61.

BARNES, A. (1977). The influence of the length of the growth period and planting density on total crop yield. *Annals of Botany,* **41**, 883–95.

BARROW, N. J. (1975). The response to phosphate of two annual pasture species. II. The specific rate of uptake of phosphate, its distribution and use for growth. *Australian Journal of Agricultural Research,* **26**, 145–56.

BARROW, N. J. (1977). Phosphorus uptake and utilization by tree seedlings. *Australian Journal of Botany*, **25**, 571–84.

BAYLY, I. L. and SHIBLEY, M. E. (1978). Seasonal nutrient and sodium accumulation in the macrophyte *Pontederia cordata. Canadian Journal of Botany*, **56**, 417–25.

BAZZAZ, F. A. and CARLSON, R. W. (1979). Photosynthetic contribution of flowers and seeds to reproductive effort of an annual colonizer. *New Phytologist*, **82**, 223–32.

BAZZAZ, F. A., CARLSON, R. W. and HARPER, J. L. (1979). Contribution to reproductive effort by photosynthesis of flowers and fruits. *Nature*, **279**, 554–5.

BAZZAZ, F. A. and HARPER, J. L. (1977). Demographic analysis of the growth of *Linum usitatissimum. New Phytologist*, **78**, 193–208.

BEALE, P. E. and THURLING, N. (1979). Genotypic variation in resistance to *Kabatiella caulivora* in *Trifolium subterraneum* subspecies *yanninicum. Australian Journal of Agricultural Research*, **30**, 651–58.

BEAN, E. W. (1971). Temperature effects upon inflorescence and seed development in Tall Fescue (*Festuca arundinacea* Schreb.). *Annals of Botany*, **35**, 891–7.

BEAN, E. W.and YOK-HWA, C. (1972). An analysis of the growth of inbred progeny of *Lolium. Journal of Agricultural Science*, **79**, 147–53.

BEARDSELL, M. F. (1977). Effects of routine handling on maize growth. *Australian Journal of Plant Physiology*, **4**, 857–61.

BERNARD, J. and NEVILLE, P. (1978). Variations de la vitesse de fonctionnement des méristèmes caulinaires de Pisum sativum liées au passage de l'état végétatif a l'état reproducteur. *Annales des Sciences Naturelles, Botanique, Paris, 12e Serie*, **19**, 283–99.

BHAT, K. K. S., BRERETON, A. J. and NYE, P. H. (1979a). The possibility of predicting solute uptake and plant growth response from independently measured soil and plant characteristics. VIII. A comparison of the growth and nitrate uptake of rape grown in similar nitrate concentration in of the results with model predictions. *Plant and Soil*, **53**, 169–91.

BHAT, K. K. S., BRERETON, A. J. and NYE, P. H. (1979b). The possibility of predicting solute uptake and plant growth response from independently measured soil and plant characteristics. VIII. A comparison of the growth and nitrate uptake of rape grown in similar nitrate concentration in solution or in soil solution. *Plant and Soil*, **53**, 193–201.

BHAT, K. K. S., NYE, P. H. and BRERETON, A. J. (1979). The possibility of predicting solute uptake and plant growth response from independently measured soil and plant characteristics. VI. The growth and uptake of rape in solutions of constant nitrate concentration. *Plant and Soil*, **53**, 137–67.

BLACK, J. N. (1963). The inter-relationship of solar radiation and leaf area index in determining the rate of dry matter production of swards of subterranean clover (*Trifolium subterraneum* L.). *Australian Journal of Agricultural Research*, **14**, 20–38.

BLACKLOW, W. M. and McGUIRE, W. S. (1971) Influence of gibberellic acid on the winter growth of varieties of tall fescue (*Festuca arundinacea* Schreb.). *Crop Science,* **11**, 19–22.

BLACKMAN, G. E. (1961). Responses to environmental factors by plants in the vegetative phase. In *Growth in Living Systems.* Ed. M. X. ZARROW, pp. 525–6. Basic Books, New York.

BLACKMAN, G. E. (1968). The application of the concepts of growth analysis to the assessment of productivity. In *Functioning of Terrestrial Ecosystems at the Primary Production Level* (Proceedings of the Copenhagen Symposium). Ed. F. E. ECKARDT, pp. 243–59. UNESCO, Paris.

BLACKMAN, V. H. and WILSON, G. L. (1951). Physiological and ecological studies in the analysis of plant environment. VII. An analysis of the differential effects of light intensity on the net assimilation rate, leaf-area ratio, and relative growth rate of differential species. *Annals of Botany,* **15**, 373–409.

BLACKMAN, V. H. (1919). The compound interest law and plant growth. *Annals of Botany,* **33**, 353–60.

BLUM, U. and HECK, W. W. (1980). Effects of acute ozone exposures on snap bean at various stages of its life cycle. *Environmental and Experimental Botany,* **20**, 73–85.

BOND, T. E. T. (1945). Studies in the vegetative growth and anatomy of the tea plant (*Camellia thea* Link.) with special reference to the phloem. *Annals of Botany,* **9**, 183–216.

BOOTE, K. J., GALLAHER, R. N., ROBERTSON, W. K., HINSON, K, and HAMMOND, L. C. (1978). Effect of foliar fertilization on photosynthesis, leaf nutrition, and yield of soybeans. *Agronomy Journal,* **70**, 787–91.

BREMNER, P. M., EL SAEED, E. A. K. and SCOTT, R. K. (1967). Some aspects of competition for light in potatoes and sugar beet. *Journal of Agricultural Science,* **69**, 283–90.

BREWSTER, J. L. (1979). The response of growth rate to temperature in seedlings of several *Allium* crop species. *Annals of Applied Biology,* **93**, 351–7.

BREWSTER, J. L., BHAT, K. K. S. and NYE, P. H. (1975). The possibility of predicting solute uptake and plant growth response from independently measured soil and plant characteristics. II. The growth and uptake of onions in solutions of constant phosphate concentration. *Plant and Soil,* **42**, 171–95.

BREWSTER, J. L. and TINKER, P. B. (1970). Nutrient cation flows in soil around plant roots. *Soil Science Society of America Proceedings,* **34**, 421–6.

BRIGGS, G. E., KIDD, F. and WEST, C. (1920). A quantitative analysis of plant growth. II. *Annals of Applied Biology,* **7**, 202–23.

BRITZ, S. J. , PFAU, J., NULTSCH, W. and BRIGGS, W. R. (1976). Automatic monitoring of a circadian rhythm of change in light transmittance in *Ulva. Plant Physiology,* **58**, 17–21.

BROUGHAM, R. W. (1955). A study in rate of pasture growth. *Australian Journal of Agricultural Research,* **6**, 804–12.

BROUGHAM, R. W. (1956). The rate of growth of short-rotation-ryegrass pastures in the late autumn, winter, and early spring. *New Zealand Journal of Science and Technology, Sect. A,* **38**, 78–87.

BUNTING, E. S. (1972). Ripening in maize: inter-relationships between time, water content and weight of dry material in ripening grain of a flint × dent hybrid (Inra 200). *Journal of Agricultural Science,* **79**, 225–33.

BURDON, J. J. and HARPER, J. L. (1980). Relative growth rates of individual members of a plant population. *Journal of Ecology,* **68**, 953–7.

BUSEY, P. and MYERS, B. J. (1979). Growth rates of turfgrass propagated vegetatively. *Agronomy Journal,* **71**, 817–21.

BUTTERY, B. R. (1969). Analysis of the growth of soybeans as affected by plant population and fertilizer. *Canadian Journal of Plant Science,* **49**, 675–84.

BUTTERY, B. R. and BUZZELL, R. I. (1974). Evaluation of methods used in computing net assimilation rates of soybeans (*Glycine max* (L.) Merrill). *Crop Science,* **14**, 41–4.

CALLAGHAN, T. V. (1974). Intraspecific variation in *Phleum alpinum* L. with specific reference to polar populations. *Arctic and Alpine Research,* **6**, 361–401.

CALLAGHAN, T. V. (1976). Growth and population dynamics of *Carex bigelowii* in an alpine environment. *Oikos,* **27**, 402–13.

CALOIN, M., EL KHODRE, A. and AIRY, M. (1980). Effect of nitrate concentration in the root:shoot ratio in *Dactylis glomerata* L. and on the kinetics of growth in the vegetative phase. *Annals of Botany,* **46**, 165–73.

CALOW, P. (1976). *Biological Machines: a Cybernetic Approach to Life.* Edward Arnold, London.

CAMPBELL, C. L., PENNYPACKER, S. P. and MADDEN, L. V. (1980). Progression dynamics of hypocotyl rot of snapbean. *Phytopathology,* **70**, 487–94.

CANNELL, M. G. R. and CAHALAN, C. M. (1979). Shoot apical meristems of *Picea sitchensis* seedlings accelerate in growth following bud-set. *Annals of Botany,* **44**, 209–14.

CANNELL, M. G. R. and WILLETT, S. C. (1976). Shoot growth phenology, dry matter distribution and root:shoot ratios of provenances of *Populus trichocarpa, Picea sitchensis* and *Pinus contorta* growing in Scotland. *Silvae Genetica,* **25**, 27–88.

CAUSTON, D. R. (1967). *Some Mathematical Properties of Growth Curves and Applications to Plant Growth Analysis.* Ph.D. Thesis, University of Wales, Aberystwyth.

CAUSTON, D. R. (1969). A computer program for fitting the Richards function. *Biometrics,* **25**, 401–9.

CAUSTON, D. R. (1970). *Growth Functions, Growth Analysis, and Plant Physiology.* Forest Commission Research Division, Statistics Section Paper No. 151, Farnham, Surrey.

CAUSTON, D. R. (1977). *A Biologist's Mathematics.* Edward Arnold, London.

CAUSTON, D. R., ELIAS, C. O. and HADLEY, P. (1978). Biometrical studies of plant growth. I. The Richards function, and its application in analysing the effects of temperature on leaf growth. *Plant, Cell and Environment,* **1**, 163–84.

CAUSTON, D. R. and MER, C. L. (1966). Analytical studies of the growth of the etiolated seedling of *Avena sativa.* I. Meristematic activity in the mesocotyl with special reference to the effect of carbon dioxide. *New Phytologist,* **65**, 87–99.

CAUSTON, D. R. and VENUS, J. C. (1981). *The Biometry of Plant Growth.* Edward Arnold, London.

CHANCE, J. E. and FOERSTER, C. O. (1973). The growth curve for an avocado graft. *Journal of the Rio Grande Valley Horticultural Society,* **27**, 71–3.

CHARLES-EDWARDS, D. A. and FISHER, M. J. (1981). A physiological approach to the analysis of crop growth data. I. Theoretical considerations. *Annals of Botany,* **46**, 413–23.

CHATTERTON, N. J. and SILVIUS, J. E. (1979). Photosynthate partitioning into starch in soybean leaves. I. Effects of photoperiod *versus* photosynthetic period duration. *Plant Physiology,* **64**, 749–53.

CHOW, C. S. and TAN, H. (1979). Variation in stimulation response in yield of a *Hevea* clone. II. A regression model. *Journal of the Rubber Research Institute of Malaysia,* **27**, 8–23.

CHRISTIANSEN, M. N. (1962). A method of measuring and expressing epigeous seedling growth rate. *Crop Science,* **2**, 487–9.

CHRISTIE, E. K. (1978). Ecosystem processes in semiarid grasslands. I. Primary production and water use of two communities possessing different photosynthetic pathways. *Australian Journal of Agricultural Research,* **29**, 773–87.

CHRISTIE, E. K. (1979). Ecosystem processes in semiarid grasslands. II. Litter production, decomposition and nutrient dynamics. *Australian Journal of Agricultural Research,* **30**, 29–42.

CHRISTIE, E. K. and MOORBY, J. (1975). Physiological responses of semiarid grasses. I. The influence of phosphorus supply on growth and phosphorus absorption. *Australian Journal of Agricultural Research,* **26**, 423–36.

CHRISTY, A. L. and FISHER, D. B. (1978). Kinetics of ^{14}C-photosynthate translocation in Morning Glory vines. *Plant Physiology,* **61**, 283–90.

CLARK, W. E. Le G. and MEDAWAR, P. B. (1945). *Essays on Growth and Form presented to D'Arcy Wentworth Thompson.* Oxford University Press, London.

CLARKE, G. M. (1980). *Statistics and Experimental Design,* second edition. Edward Arnold, London.

CLARKE J. M. and SIMPSON, G. M. (1978). Growth analysis of *Brassica napus* cv. Tower. *Canadian Journal of Plant Science,* **58**, 587–95.

CLARKE, J. M. and SIMPSON, G. M. (1979). The application of a curve-fitting technique to *Brassica napus* growth data. *Field Crops Research,* **2**, 35–43.

COCK, J. H., FRANKLIN, D., SANDOVAL, G. and JURI, P. (1979). The ideal cassava plant for maximum yield. *Crop Science,* **19**, 271–9.

CONSTABLE, G. A. and GLEESON, A. C. (1977). Growth and distribution of dry matter in cotton (*Gossypium hirsutum* L.). *Australian Journal of Agricultural Research,* **28**, 249–56.

CONSTABLE, G. A. and RAWSON, H. M. (1980). Carbon production and utilization in cotton: inferences from a carbon budget. *Australian Journal of Plant Physiology,* **7**, 539–53.

COOMBE, D. E. (1960). An analysis of the growth of *Trema guineensis. Journal of Ecology,* **48**, 219–31.

COPEMAN, P. R. van den R. (1928). Autocatalysis and growth. *Annals of Applied Biology,* **15**, 613–22.

CRITTENDEN, P. D. and READ, D. J. (1978). The effects of air pollution on plant growth with special reference to sulphur dioxide. II. Growth studies with *Lolium perenne* L. *New Phytologist,* **80**, 49–62.

CRITTENDEN, P. D. and READ, D. J. (1979). The effects of air pollution on plant growth with special reference to sulphur dioxide. III. Growth studies with *Lolium multiflorum* Lam. and *Dactylis glomerata* L. *New Phytologist,* **83**, 645–51.

CRUIZIAT, P., TYREE, M. T., BODET, C. and LoGULLO, M. A. (1980). The kinetics of rehydration of detached sunflower leaves following substantial water loss. *New Phytologist,* **84**, 293–306.

DALE, R. F., COELHO, D. T. and GALLO, K. P. (1980). Prediction of daily green leaf area index for corn. *Agronomy Journal,* **72**, 999–1005.

DATA, E. S. and PRATT, H. K. (1980). Patterns of pod growth, development, and respiration in the winged bean (*Psophocarpus tetragonolobus*). *Tropical Agriculture (Trinidad),* **57**, 309–17.

DAVIES, A. (1971). Growth rates and crop morphology in vernalized and non-vernalized swards of perennial ryegrass in spring. *Journal of Agricultural Science,* **77**, 273–82.

DAVIES, O. L. and KU, J. Y. (1977). A re-examination of the fitting of the Richards growth function. *Biometrics,* **33**, 546–7.

DENNETT, M. D. and AULD, B. A. (1980). The effects of position and temperature on the expansion of leaves of *Vicia faba* L. *Annals of Botany,* **46**, 511–7.

DENNETT, M. D., AULD, B. A. and ELSTON, J. (1978). A description of leaf growth in *Vicia faba* L. *Annals of Botany,* **42**, 223–32.

DENNETT, M. D., ELSTON, J. and MILFORD, J. R. (1979). The effect of temperature on the growth of individual leaves of *Vicia faba* L. in the field. *Annals of Botany,* **43**, 197–208.

DE VRIES, D. P. (1976). Juvenility in hybrid tea roses. *Euphytica,* **25**, 321–8.

DE VRIES, D. P. and DUBOIS, L. A. M. (1977). Early selection in hybrid tea-rose seedlings for cut stem length. *Euphytica,* **26**, 761–4.

DIXON, W. J. (Ed.) (1973). *BMD Biomedical Computer Programs.* University of California Press, Berkley and Los Angeles.

DOLEY, D. (1978). Effects of shade on gas exchange and growth in seedlings of *Eucalyptus grandis* Hill ex Maiden. *Australian Journal of Plant Physiology,* **5**, 723–38.

DRAPER, N. R. and SMITH, H. (1966). *Applied Regression Analysis.* John Wiley, New York.

DuCHATEAU, P. C., NOFZIGER, D. L., AHUJA, L. R. and SWARTZENDRUBER, D. (1972). Experimental curves and rates of change from piecewise parabolic fits. *Agronomy Journal,* **64**, 538–42.

DUDNEY, P. J. (1973). An approach to the growth analysis of perennial plants. *Proceedings of the Royal Society of London, Series B,* **184**, 217–20.

DUDNEY, P. J. (1974). An analysis of growth rates in the early life of apple trees. *Annals of Botany,* **38**, 647–56.

DUNCAN, D. B. (1955). Multiple range and multiple *F* tests. *Biometrics,* **11**, 1–42.

DUNCAN, W. G. and HESKETH, J. D. (1968). Net photosynthetic rates, relative leaf growth rates, and leaf numbers of 22 races of maize grown at eight temperatures. *Crop Science,* **8**, 670–4.

DYKYJOVÁ, D., ONDOK, J. P. and ŘÍBÁŇ, K. (1970). Seasonal changes in productivity and vertical structure of reed-stands (*Phragmites communis* Trin.). *Photosynthetica,* **4**, 280–7.

EAGLES, C. F. (1969). Time changes of relative growth-rate in two natural populations of *Dactylis glomerata* L. *Annals of Botany,* **33**, 937–46.

EAGLES, C. F. (1971). Changes in net assimilation rate and leaf-area ratio with time in *Dactylis glomerata* L. *Annals of Botany,* **35**, 63–74.

ECKARDT, F. E., HEIM, G., METHY, M., SAUGIER, F. and SAUVEZON, R. (1971). Fonctionnement d'un écosystème au niveau de la production primaire mesures effectuées dans une culture d'*Helianthus annuus. Oecologia Plantarum,* **6**, 51–100.

EDELSTEN, P. R. and CORRALL, A. J. (1979). Regression models to predict herbage production and digestibility in a non-regular sequence of cuts. *Journal of Agricultural Science,* **92**, 575–85.

EHRLICH, P. R. and EHRLICH, A. H. (1972). *Population, Resources, Environment,* second edition. W. H. Freeman, San Francisco.

ELIAS, C. O. and CAUSTON, D. R. (1976). Studies on data variability and the use of polynomials to describe plant growth. *New Phytologist,* **77**, 421–30.

ELIAS, C. O. and CHADWICK, M. J. (1979). Growth characteristics of grass and legume cultivars and their potential for land reclamation. *Journal of Applied Ecology,* **16**, 537–44.

ELLIOTT, B. R. and JARDINE, R. (1972). The influence of rotation systems on long-term trends in wheat yield. *Australian Journal of Agricultural Research,* **23**, 935–44.

ELLIOTT, J. R. and PEIRSON, D. R. (1980). A response surface analysis of the effects of cyclohexanecarboxylic acid and 2,4-dichlorophenoxy acetic acid on nitrogen metabolism in *Phaseolus vulgaris* L. *Annals of Botany,* **46**, 577–91.

EL LOZY, M. (1978). A critical analysis of the double and triple logistic growth curves. *Annals of Human Biology*, **5**, 389–94.

ELMORE, C. D., HESKETH, J. D. and MURAMOTO, H. (1967). A survey of rates of leaf growth, leaf aging and leaf photosynthetic rates among and within species. *Journal of the Arizona Academy of Science*, **4**, 215–9.

EMECZ, T. I. (1962). Suggested amendments in growth analysis and potentiality assessment in relation to light. *Annals of Botany*, **26**, 517–27.

ENGLISH, S. D., McWILLIAM, J. R., SMITH, R. C. G. and DAVIDSON, J. L. (1979). Photosynthesis and partitioning of dry matter in sunflower. *Australian Journal of Plant Physiology*, **6**, 149–64.

ERDÖS, L. (1980). Changes in the yield structure of maize. *Acta Agronomica Academiae Scientiarum Hungaricae*, **29**, 50–62.

ERH, K. T. (1972). Application of the spline function to soil science. *Soil Science*, **114**, 333–8.

ERICKSON, R. O. (1959). Integration of plant growth processes. *The American Naturalist*, **93**, 225–35.

ERICKSON, R. O. (1976). Modelling of plant growth. *Annual Review of Plant Physiology*, **27**, 407–34.

ERICKSON, R. O. and SAX, K. B. (1956a). Elemental growth rate of the primary root of *Zea mays*. *Proceedings of the American Philosophical Society*, **100**, 487–98.

ERICKSON, R. O. and SAX, K. B. (1956b). Rates of cell division and cell elongation in the growth of the primary root of *Zea mays*. *Proceedings of the American Philosophical Society*, **100**, 499–514.

EVANS, G. C. (1972). *The Quantitative Analysis of Plant Growth*. Blackwell Scientific Publications, Oxford.

EVANS, G. C. (1976). A sack of uncut diamonds: the study of ecosystems and the future resources of mankind. *Journal of Ecology*, **64**, 1–39.

EVANS, G. C. and HUGHES, A. P. (1962). Plant growth and the aerial environment. III. On the computation of unit leaf rate. *New Phytologist*, **61**, 322–7.

FARRAR, J. F. (1978). Ecological physiology of the lichen *Hypogymnia physodes*. IV. Carbon allocation at low temperatures. *New Phytologist*, **81**, 65–9.

FARRAR, J. F. (1980). Allocation of carbon to growth, storage and respiration in the vegetative barley plant. *Plant, Cell and Environment*, **3**, 97–105.

FERRARIS, R. and SINCLAIR, D. F. (1980). Factors affecting the growth of *Pennisetum purpureum* in the wet tropics. II. Uninterrupted growth. *Australian Journal of Agricultural Research*, **31**, 915–25.

FINNEY, D. J. (1978). Statistics and statisticians in agricultural research. *Journal of Agricultural Science*, **91**, 653–9.

FISCHER, K. S. and WILSON, G. L. (1975). Studies of grain production in *Sorghum bicolor* (L. Moench.). V. Effect of planting density on growth and yield. *Australian Journal of Agricultural Research*, **26**, 31–41.

FISHER, N. M. and MILBOURN, G. M. (1974). The effect of plant density, date of apical bud removal and leaf removal on the growth and yield of single-harvest Brussels sprouts (*Brassica oleracea* var. *gemmifera* D.C.). 1. Whole plant and axillary bud growth. *Journal of Agricultural Science*, **83**, 479–87.

FISHER, R. A. (1921). Some remarks on the methods formulated in a recent article on 'The quantitative analysis of plant growth'. *Annals of Applied Biology*, **7**, 367–72.

FISHER, R. A. (1966). *The Design of Experiments*, eighth edition. Oliver and Boyd, Edinburgh and London.

FISHER, R. A. (1970). *Statistical Methods for Research Workers*, fourteenth edition prepared by E. A. CORNISH. Oliver and Boyd, Edinburgh.

FISHER, R. A. and YATES, F. (1963). *Statistical Tables for Biological, Agricultural and Medical Research*, sixth edition. Oliver and Boyd, Edinburgh.

FITZHUGH, H. A. Jr. (1976). Analysis of growth curves and strategies for altering their shape. *Journal of Animal Science*, **42**, 1036–51.

FONDY, B. R. and GEIGER. D. R. (1980). Effect of rapid changes in sink-source ratio on export and distribution of products of photosynthesis in leaves of *Beta vulgaris* L. and *Phaseolus vulgaris* L. *Plant Physiology*, **66**, 945–9.

FRALEY, L. Jr. and WHICKER, F. W. (1973a). Response of shortgrass plains vegetation to gamma radiation. I. Chronic irradiation. *Radiation Botany*, **13**, 331–41.

FRALEY, L. Jr. and WHICKER, F. W. (1973b). Response of shortgrass plains vegetation to gamma radiation. II. Short-term seasonal irradiation. *Radiation Botany*, **13**, 343–53.

FRESCO, L. F. M. (1973). A model for plant growth. Estimation of the parameters of the logistic function. *Acta Botanica Neerlandica*, **22**, 486–9.

FREYMAN, S. (1980). Quantitative analysis of growth in Southern Alberta of two barley cultivars grown from magnetically treated and untreated seed. *Canadian Journal of Plant Science*, **60**, 463–71.

FRIBOURG, H. A., BARTH, K. M., MCLAREN, J. B., CARVER, L. A., CONNELL, J. T. and BRYAN, J. M. (1979). Seasonal trends of in vitro day matter digestibility of N-fertilized bermudagrass and of orchardgrass-ladino pastures. *Agronomy Journal*, **71**, 117–20.

FRIEDRICH, J. W. and SCHRADER, L. E. (1979). N deprivation in maize during grain-filling. II. Remobilization of ^{15}N and ^{35}S and the relationship between N and S accumulation. *Agronomy Journal*, **71**, 466–72.

FREIDRICH, J. W., SCHRADER, L. E. and NORDHEIM, E. V. (1979). N deprivation in maize during grain-filling. I. Accumulation of dry matter, nitrate-N, and sulfate-S. *Agronomy Journal*, **71**, 461–5.

FRIEND, D. J. C. (1960). The control of chlorophyll accumulation in leaves of Marquis wheat by temperature and light intensity. I. The rate of chlorophyll accumulation and maximal absolute chlorophyll contents. *Physiologia Plantarum*, **13**, 776–85.

FRIEND, D. J. C., HELSON, V. A. and FISHER, J. E. (1962). The rate of dry weight accumulation in Marquis wheat as affected by temperature and light intensity. *Canadian Journal of Botany,* **40,** 939–55.

FRIEND, D. J. C., HELSON, V. A. and FISHER, J. E. (1965). Changes in the leaf area ratio during growth of Marquis wheat, as affected by temperature and light intensity. *Canadian Journal of Botany,* **43,** 15–28.

FUKAI, S. and SILSBURY, J. H. (1976). Responses of subterranean clover communities to temperature. I. Dry matter production and plant morphogenesis. *Australian Journal of Plant Physiology,* **3,** 527–43.

FUKAI, S. and SILSBURY, J. H. (1977). Responses of subterranean clover communities to temperature. II. Effects of temperature on dark respiration rate. *Australian Journal of Plant Physiology,* **4,** 159–67.

FULLER, W. A. (1969). Grafted polynomials as approximating functions. *Australian Journal of Agricultural Economics,* **13,** 35–46.

GAINES, W. L. and NEVENS, W. B. (1925). Growth-equation constants in crop studies. *Journal of Agricultural Science,* **31,** 973–85.

GANDAR, P. W. (1980). The analysis of growth and cell production in root apices. *Botanical Gazette,* **141,** 131–8.

GAUSS, C. F. (1809). *Theoria motus corporum coelestium in sectionibus conicis solem ambientium* (article 115). Reprinted in: *Carl Friedrich Gauss Werke* VII. (Ed. E. J. SCHERING). Perthes, Gotha 1871.

GERAKIS, P. A. and PAPAKOSTA-TASOPOULOU, D. (1979). Growth dynamics of *Zea mays* L. populations differing in genotype and density and grown under illuminance stress. *Oecologia Plantarum,* **14,** 13–26.

GLADILIN, K. L. and ORLOVSKII, A. F. (1973). Method of selection of approximating functions. *Doklady Akademii Nauk SSR,* **211,** 230–3.

GLENDAY, A. C. (1955). The mathematical separation of plant and weather effects in field growth studies. *Australian Journal of Agricultural Research,* **6,** 813–22.

GLENDAY, A. C. (1959). Mathematical analysis of growth curves replicated in time. *New Zealand Journal of Agricultural Research,* **2,** 297–305.

GLOAGUEN, J. C. and TOUFFET, J. (1980). Vitesse de décomposition et évolution minérale des litières sous climat atlantique. *Acta Oecoligia/ Oecologia Plantarum,* **1,** 3–26.

GLOAGUEN, J. C., TOUFFET, J. and FORGEARD, F. (1980). Vitesse de décomposition et évolution minérale des litières sous climat atlantique. II. Les principales especes des landes de Bretagne (France). *Acta Oecologia/ Oecologia Plantarum,* **1,** 257–73.

GOLDSWORTHY, P. R. (1970). The growth and yield of tall and short sorghums in Nigeria. *Journal of Agricultural Science,* **75,** 109–22.

GOLDSWORTHY, P. R. and COLEGROVE, M. (1974). Growth and yield of highland maize in Mexico. *Journal of Agricultural Science,* **83,** 213–21.

GOLDSWORTHY, P. R., PALMER, A. F. E. and SPERLING, D. W. (1974). Growth and yield of lowland tropical maize in Mexico. *Journal of Agricultural Science,* **83,** 223–30.

GOODALL, D. W. (1945). The distribution of weight change in the young tomato plant. I. Dry-weight changes of the various organs. *Annals of Botany*, **9**, 101–39.

GOODALL, D. W. (1949). A quantitative study of the early development of the seedling of cacao (*Theobroma cacao*). *Annals of Botany*, **13**, 1–21.

GOODALL, D. W. (1950). Growth analysis of cacao seedlings. *Annals of Botany*, **14**, 291–306.

GOODMAN, P. G. (1968). Physiological analysis of the effects of different soils on sugar beet crops in different years. *Journal of Applied Ecology*, **5**, 339–57.

GORDON, I. L. (1979). Selection against sprouting damage in wheat. III. Dormancy, germinative alpha-amylase, grain redness and flavanols. *Australian Journal of Agricultural Research*, **30**, 387–402.

GORDON, I. L., BALAAM, L. N. and DERERA, N. F. (1977). Selection against sprouting damage in wheat. II. Harvest ripeness, grain maturity and germinability. *Australian Journal of Agricultural Research*, **28**, 583–99.

GRACE, J. and WOOLHOUSE, H. W. (1970). A physiological and mathematical study of the growth and productivity of a *Calluna-Sphagnum* community. I. Net photosynthesis of *Calluna vulgaris* L. Hull. *Journal of Applied Ecology*, **7**, 363–81.

GRACE, J. and WOOLHOUSE, H. W. (1973). A physiological and mathematical study of the growth and productivity of a *Calluna-Sphagnum* community. III. Distribution of photosynthate in *Calluna vulgaris* L. Hull. *Journal of Applied Ecology*, **10**, 77–91.

GRAVES, C. J. (1978). Uptake and distribution of copper in *Chrysanthemum morifolium*. *Annals of Botany*, **42**, 117–25.

GRAY, D. and MORRIS, G. E. L. (1978). Seasonal effects on the growth and time to maturity of lettuce. *Journal of Agricultural Science*, **91**, 523–9.

GREGORY, F. G. (1918). Physiological conditions in cucumber houses. *3rd Annual Report of the Experimental Research Station, Cheshunt*, 19–28.

GREGORY, F. G. (1921). Studies in the energy relations of plants. I. The increase in area of leaves and leaf surfaces of *Cucumis sativus*. *Annals of Botany*, **35**, 93–123.

GREGORY, F. G. (1926). The effect of climatic conditions on the growth of barley. *Annals of Botany*, **40**, 1–26.

GREGORY, F. G. (1928a). The analysis of growth curves. A reply to criticism. *Annals of Botany*, **42**, 531–9.

GREGORY, F. G. (1928b). Studies in the energy relations of plants. II. The effect of temperature on increase in area of leaf surface and in dry weight of *Cucumis sativus*. Part I. The effect of temperature on the increase in area of leaf surface. *Annals of Botany*, **42**, 469–507.

GREVILLE, T. N. E. (Ed.) (1969). *Theory and Applications of Spline Functions*. Academic Press, New York.

GRIGGS, M. M., NANCE, W. L. and DINUS, R. J. (1978). Analysis and comparison of fusiform rust disease progress curves for five slash pine families. *Phytopathology*, **68**, 1631–6.

GRIME, J. P. (1974). Vegetation classification by reference to strategies. *Nature, London,* **250**, 26–31.

GRIME, J. P. (1977). Evidence for the existence of three primary strategies in plants and its relevance to ecological and evolutionary theory. *American Naturalist,* **111**, 1169–94.

GRIME, J. P. (1979). *Plant Strategies and Vegetation Processes.* John Wiley and Sons, Chichester.

GRIME, J. P. and HUNT, R. (1975). Relative growth-rate: its range and adaptive significance in a local flora. *Journal of Ecology,* **63**, 393–422.

GRIMM, H. (1977). On growth models and analysis of growth curves in microbiology. *Biometrical Journal,* **19**, 529–34.

GRUNOW, J. O., GROENEVELD, H. T. and DU TOIT, S. H. C. (1980). Above-ground dry matter dynamics of the grass layer of a South African tree savanna. *Journal of Ecology,* **68**, 877–89.

HACKETT, C. (1969). A study of the root system of barley. II. Relationships between root dimensions and nutrient uptake. *New Phytologist,* **68**, 1023–30.

HACKETT, C. and RAWSON, H. M. (1974). An exploration of the carbon economy of the tobacco plant. II. Patterns of leaf growth and dry matter partitioning. *Australian Journal of Plant Physiology,* **1**, 271–81.

HADLEY, P. (1978). *Growth and Photosynthetic Activity of Individual Leaves of Plants during Growth.* Ph.D. Thesis, University of Wales.

HADLEY, P., BOXALL, M. I., RICHARDSON, A. C., DICKINSON, D., MINCHIN, F. R., SUMMERFIELD, R. J. and ROBERTS, E. H. (1980). A system for continuous monitoring of whole shoot CO_2 exchange as an adjunct to growth analysis experiments in controlled environments. *Journal of Experimental Botany,* **31**, 679–89.

HALL, A. J. (1977). Assimilate source-sink relationships in *Capsicum annuum* L. I. The dynamics of growth in fruiting and deflorated plants. *Australian Journal of Plant Physiology,* **4**, 623–36.

HAMILTON, G. J. (1975). *Forest Mensuration Handbook.* Forestry Commission Booklet no. 39. H.M.S.O., London.

HAMMERTON, J. L. and STONE, M. (1966). Studies on weed species of the genus *Polygonum* L. II. Physiological variation within *P. lapathifolium* L. *Weed Research,* **6**, 104–31.

HAMMOND, L. C. and KIRKHAM, D. (1949). Growth curves of soybeans and corn. *Agronomy Journal,* **41**, 23–9.

HANNAM, R. V. (1968). Leaf growth and development in the young tobacco plant. *Australian Journal of Biological Sciences,* **21**, 855–70.

HARPER, J. L. (1977). *Population Biology of Plants.* Academic Press, London.

HARPER, J. L. and WHITE, J. (1974). The demography of plants. *Annual Review of Ecology and Systematics,* **5**, 419–63.

HASEGAWA, K. (1976). Study on the method in the comparison of height growth curves – in the case of Sugi (*Cryptomeria japonica*) in the south-west of Honshu Island. *Bulletin of the Government Forestry Experimental Station,* **286**, 53–74.

HEARN, A. B. (1969a). Growth and performance of cotton in a desert environment. I. Morphological development of the crop. *Journal of Agricultural Science,* **73**, 65–74.

HEARN, A. B. (1969b). The growth and performance of cotton in a desert environment. II. Dry matter production. *Journal of Agricultural Science,* **73**, 75–86.

HEARN, A. B. (1972a). The growth and performance of rain-grown cotton in a tropical upland environment. I. Yields, water relations and crop growth. *Journal of Agricultural Science,* **79**, 121–35.

HEARN, A. B. (1972b). The growth and performance of rain-grown cotton in a tropical upland environment. II. The relationship between yield and growth. *Journal of Agricultural Science,* **79**, 137–45.

HEATH, O. V. S. (1932). Growth curves, and co-ordination between Stations. *Empire Cotton Growing Corporation, Reports from Experimental Stations 1930–31,* 28–34.

HEATH, O. V. S. (1937a). The effect of age on net assimilation and relative growth rates in the cotton plant. *Annals of Botany,* **1**, 565–6.

HEATH, O. V. S. (1937b). The growth in height and weight of the cotton plant under field conditions. *Annals of Botany,* **1**, 515–20.

HEATH, O. V. S. (1970). *Investigation by Experiment.* Studies in Biology No. 23. Edward Arnold, London.

HEDLEY, C. L. and AMBROSE, M. J. (1980). An analysis of seed development in *Pisum sativum* L. *Annals of Botany,* **46**, 89–105.

HERRERA, R. S. and RAMOS, N. (1980). Bermuda grass response to nitrogen fertilization and age of regrowth. I. Mineral composition in the dry season. *Cuban Journal of Agricultural Science,* **14**, 79–86.

HEWLETT, P. S. and PLACKETT, R. L. (1979). *An Introduction to the Interpretation of Quantal Responses in Biology.* Edward Arnold, London.

HICKS, P. A. and ASHBY, E. (1934). Interaction of factors in the growth of *Lemna.* V. Some preliminary observations upon the interaction of temperature and light on the growth of *Lemna. Annals of Botany,* **48**, 515–25.

HIRSCHFELD, W. J. (1970a). Time series and exponential smoothing methods applied to the analysis and prediction of growth. *Growth,* **34**, 129–43.

HIRSCHFELD, W. J. (1970b). A comparison of regression with time series-exponential smoothing predictions of craniofacial growth. *Growth,* **34**, 431–5.

HO, L. C. and SHAW, L. F. (1979). Net accumulation of minerals and water and the carbon budget in an expanding leaf of tomato. *Annals of Botany,* **43**, 45–54.

HODGKINSON, K. C. and QUINN, J. A. (1976). Adaptive variability in the growth of *Danthonia caespitosa* Gaud. Populations at different temperatures. *Australian Journal of Botany,* **24**, 381–96.

HOGETSU, K. OSHIMA, Y., MIDORIKAWA, B., TEZUKA, Y., SAKAMOTO, M., MOTOTANI I. and KIMURA, M. (1960). Growth analytical studies on the artificial communities of *Helianthus tuberosus* with different densities. *Japanese Journal of Botany,* **17**, 278–305.

HOLLIDAY, R. J. and PUTWAIN, P. D. (1980). Evolution of herbicide resistance in *Senecio vulgaris:* variation in susceptibility to simazine between and within populations. *Journal of Applied Ecology,* **17**, 779–91.

HOMÈS, M. V. and VAN SCHOOR, G. H. (1978). Les courbes de croissance des végétaux et leur modification sous l'effet de l'alimentation minérale. *Bulletin de la Classe des Sciences Academie Royale de Belgique, Series 5,* **64**, 425–39.

HORSMAN, D. C., NICHOLLS, A. O. and CALDER, D. M. (1980). Growth responses of *Dactylis glomerata, Lolium perenne* and *Phalaris aquatica* to chronic ozone exposure. *Australian Journal of Plant Physiology,* **7**, 511–7.

HUDSON, D. J. (1966). Fitting segmented curves whose join points have to be estimated. *Journal of the American Statistical Association,* **61**, 1097–1129.

HUETT, D. O. and O'NEILL, G. H. (1976). Growth and development of short and long season sweet potatoes in sub-tropical Australia. *Experimental Agriculture,* **12**, 385–94.

HUGHES, A. P. (1965). Plant growth and the aerial environment. IX. A synopsis of the autecology of *Impatiens parviflora. New Phytologist,* **64**, 399–413.

HUGHES, A. P. (1969). Mutual shading in quantitative studies. *Annals of Botany,* **33**, 381–8.

HUGHES, A. P. (1973a). A comparison of the effects of light intensity and duration on *Chrysanthemum morifolium* cv. Bright Golden Anne in controlled environments. I. Growth analysis. *Annals of Botany,* **37**, 267–74.

HUGHES, A. P. (1973b). A comparison of the effects of light intensity and duration on *Chrysanthemum morifolium* cv. Bright Golden Anne in controlled environments. II. Ontogenetic changes in respiration. *Annals of Botany,* **37**, 275–86.

HUGHES, A. P. and COCKSHULL, K. E. (1969). Effects of carbon dioxide concentration on the growth of *Callistephus chinensis* cultivar Johannistag. *Annals of Botany,* **33**, 351–65.

HUGHES, A. P. and COCKSHULL, K. E. (1971a). The effects of light intensity and carbon dioxide concentration on the growth of *Chrysanthemum morifolium* cv. Bright Golden Anne. *Annals of Botany.* **35**, 899–914.

HUGHES, A. P. and COCKSHULL, K. E. (1971b). A comparison of the effects of diurnal variation in light intensity with constant light intensity on growth of *Chrysanthemum morifolium* cv. Bright Golden Anne. *Annals of Botany,* **35**, 927–32.

HUGHES, A. P. and COCKSHULL, K. E. (1971c). The variation in response to light intensity and carbon dioxide concentration shown by two cultivars of *Chrysanthemum morifolium* grown in controlled environments at two times of year. *Annals of Botany,* **35**, 933–45.

HUGHES, A. P. and COCKSHULL, K. E. (1972). Further effects of light intensity, carbon dioxide concentration, and day temperature on the growth of *Chrysanthemum morifolium* cv. Bright Golden Anne in controlled environments. *Annals of Botany,* **36**, 533–50.

HUGHES, A. P. and EVANS, G. C. (1962). Plant growth and the aerial environment. II. Effects of light intensity on *Impatiens parviflora*. *New Phytologist*, **61**, 154–74.

HUGHES, A. P. and FREEMAN, P. R. (1967). Growth analysis using frequent small harvests. *Journal of Applied Ecology*, **4**, 553–60.

HULL, C. H. and NIE, N. H. (Eds) (1981). *SPSS Update 7–9: New Procedures and Facilities for Releases 7–9.* McGraw-Hill, New York.

HUMPHRIES, E. C. (1968). The effect of growth regulators, CCC and B9, on protein and total nitrogen of bean leaves (*Phaseolus vulgaris*) during development. *Annals of Botany*, **32**, 497–507.

HUNT, R. (1973). A method of estimating root efficiency. *Journal of Applied Ecology*, **10**, 157–64.

HUNT, R. (1975). Further observations on root-shoot equilibria in perennial ryegrass (*Lolium perenne* L.). *Annals of Botany*, **39**, 745–55.

HUNT, R. (1978a). *Plant Growth Analysis.* Studies in Biology No. 96. Edward Arnold, London.

HUNT, R. (1978b). Demography versus plant growth analysis. *New Phytologist*, **80**, 269–72.

HUNT, R. (1979). Plant growth analysis: the rationale behind the use of the fitted mathematical function. *Annals of Botany*, **43**, 245–9.

HUNT, R. (1980). Diurnal patterns of dry weight increment and short-term plant growth studies. *Plant, Cell and Environment*, **3**, 475–8.

HUNT, R. (1981). The fitted curve in plant growth studies. In *Mathematics and Plant Physiology*. Eds D. A. ROSE and D. A. CHARLES-EDWARDS, pp. 283–98. Academic Press, London.

HUNT, R. (In preparation). Short-term plant growth studies involving automated destructive harvesting.

HUNT, R. and BAZZAZ, F. A. (1980). The biology of *Ambrosia trifida* L. V. Response to fertilizer, with growth analysis at the organismal and sub-organismal levels. *New Phytologist*, **84**, 113–21.

HUNT, R. and BURNETT, J. A. (1973). The effects of light intensity and external potassium level on root/shoot ratio and rates of potassium uptake in perennial ryegrass (*Lothium perenne* L.). *Annals of Botany*, **37**, 519–37.

HUNT, R. and EVANS, G. C. (1980). Classical data on the growth of maize: curve fitting with statistical analysis. *New Phytologist*, **86**, 155–80.

HUNT, R. and PARSONS, I. T. (1974). A computer program for deriving growth-functions in plant growth-analysis. *Journal of Applied Biology*, **11**, 297–307.

HUNT, R. and PARSONS, I. T. (1977). Plant growth-analysis: further applications of a recent curve-fitting program. *Journal of Applied Ecology*, **14**, 965–8.

HUNT, R. and PARSONS, I. T. (1981). *Plant Growth Analysis: User's Instructions for the Stepwise and Spline Programs.* pp. iv + 48. Unit of Comparative Plant Ecology (Natural Environment Research Council), University of Sheffield.

HUNT, R., STRIBLEY, D. P. and READ, D. J. (1975). Root/shoot equilibria in cranberry (*Vaccinium macrocarpon* Ait.). *Annals of Botany,* **39**, 807–10.

HUNT, W. F. (1970). The influence of leaf death on the rate of accumulation of green herbage during pasture regrowth. *Journal of Applied Ecology,* **7**, 41–50.

HUNT, W. F. and LOOMIS, R. S. (1976). Carbohydrate-limited growth kinetics of tobacco (*Nicotiana rustica* L.), Callus. *Plant Physiology,* **57**, 802–5.

HURD, R. G. (1968). Effects of CO_2-enrichment on the growth of young tomato plants in low light. *Annals of Botany,* **32**, 531–42.

HURD, R. G. (1977). Vegetative plant growth analysis in controlled environments. *Annals of Botany,* **41**, 779–87.

HURD, R. G., GAY, A. P. and MOUNTIFIELD, A. C. (1979). The effect of partial flower removal on the relation between root, shoot and fruit growth in the indeterminate tomato. *Annals of Applied Biology,* **93**, 77–89.

HURD, R. G. and THORNLEY, J. H. M. (1974). An analysis of the growth of young tomato plants in water culture at different light integrals and CO_2 concentrations. 1. Physiological aspects. *Annals of Botany,* **38**, 375–88.

HÜSKEN, D., STEUDLE, E. and ZIMMERMANN, U. (1978). Pressure probe technique for measuring water relations of cells in higher plants. *Plant Physiology,* **61**, 158–63.

HUTCHINGS, M. J. (1975). Some statistical problems associated with determinations of population parameters for herbaceous plants in the field. *New Phytologist,* **74**, 349–63.

HZULÁK, J. and MATEJKA, F. (1980). Study of xylem pressure potential daily dynamics by means of autocorrelation analysis. *Biologia Plantarum (Praha),* **22**, 336–340.

IDRIS, H. and MILTHORPE, F. L. (1966). Light and nutrient supplies in the competition between barley and charlock. *Oecologia Plantarum,* **1**, 143–64.

INCOLL, L. D., LONG, S. P. and ASHMORE, M. R. (1977). SI units in publications in plant science. *Current Advances in Plant Science,* **28**, 331–43.

JARVIS, B. C. (1979). The influence of cotyledons on embryonic axes during induction of dormancy in *Corylus avellana. Physiologia Plantarum,* **45**, 363–6.

JARVIS, B. C. and WILSON, D. (1977). Gibberellin effects within hazel (*Corylus avellana* L.) seeds during the breaking of dormancy. I. A direct effect of gibberellin on the embryonic axis. *New Phytologist,* **78**, 397–401.

JARVIS, B. C. and WILSON, D. A. (1978). Factors influencing the growth of embryonic axes from dormant seeds of hazel (*Corylus avellana* L.). *Planta,* **138**, 189–91.

JARVIS, B. C., WILSON, D. A. and FOWLER, M. W. (1978). Growth of isolated embryonic axes from dormant seeds of hazel (*Corylus avellana* L.). *New Phytologist,* **80**, 117–23.

JARVIS, P. G. and JARVIS, M. S. (1964), Growth rates of woody plants. *Physiologia Plantarum,* **17**, 654–66.

JEFFERS, J. N. R. (1978). *Design of Experiments.* Statistical Checklist, 1. Institute of Terrestrial Ecology (NERC), Cambridge.

JEFFERS, J. N. R. (1979). *Sampling.* Statistical Checklist, 2. Institute of Terrestrial Ecology (NERC), Cambridge.

JEFFERS, J. N. R. (1980). *Modelling.* Statistical Checklist, 3. Institute of Terrestrial Ecology (NERC), Cambridge.

JEŠKO, T. (1972). Removal of all nodal roots initiating the extension growth in *Sorghum saccharatum* (L.) Moench. *Photosynthetica,* **6**, 282–90.

JOHNSON, D. H., SARGEANT, A. B. and ALLEN, S. H. (1975). Fitting Richards' curve to data of diverse origins. *Growth,* **39**, 315–30.

JONES, C. A., REEVES, A. III, SCOTT, J. D. and BROWN, D. A. (1978). Comparison of root activity in vegetative and reproductive soybean plants. *Agronomy Journal,* **70**, 751–5.

JONES, D. F. (1918). *The Effects of Inbreeding and Crossbreeding upon Development.* Connecticut Agricultural Experimental Station, Bulletin 207.

JONES, L. (1979). The effect of stage of growth on the rate of drying of cut grass at 20°C. *Grass and Forage Science,* **34**, 139–44.

JOOSENS, J. F. and BREMS-HEYNS, E. (1975). High power polynomial regression for the study of distance, velocity and acceleration of growth *Growth,* **39**, 535–51.

KALMBACHER, R. S., MINNICK, D. R. and MARTIN, F. G. (1979). Destruction of sod-seeded legume seedlings by the snail (*Polygyra cereolus*). *Agronomy Journal,* **71**, 365–8.

KAPPLEMAN, A. J. (1980). Long-term progress made by cotton breeders in developing *Fusarium* wilt resistant germplasm. *Crop Science,* **20**, 613–5.

KARLEN, D. L. and WHITNEY, D. A. (1980). Dry matter accumulation, mineral concentrations, and nutrient distribution in winter wheat. *Agronomy Journal,* **72**, 281–8.

KEAY, J., BIDDISCOMBE, E. F. and OZANNE, P. G. (1970). The comparative rates of phosphate absorption by eight annual pasture species. *Australian Journal of Agricultural Research,* **21**, 33–44.

KENDALL, M. G. (1973). *Time-Series.* Griffin, London.

KERSHAW, K. A. (1962). Quantitative ecological studies from Landmannahellir, Iceland. III. Variation of performance in *Carex bigelowii. Journal of Ecology,* **50**, 393–9.

KEYFITZ, N. (1968). *Introduction to the Mathematics of Population.* Addison-Wesley, Reading, Mass.

KHASAWNEH, F. E. (1975). Calculation of rate of nutrient uptake by growing roots. *Agronomy Journal,* **67**, 574–6.

KIDWELL, J. F., HOWARD, A. and LAIRD, A. K. (1969). The inheritance of growth and form in the mouse. II. The Gompertz growth equation. *Growth,* **33**, 339–52.

KIMBALL, B. A. (1974). Smoothing data with fourier transformations. *Agronomy Journal,* **66**, 259–62.

KIMBALL, B. A. (1976). Smoothing data with cubic splines. *Agronomy Journal,* **68**, 126–9.

KIMURA, M., YOKOI, Y. and HOGETSU, K. (1978). Quantative relationships between growth and respiration. II. Evaluation of constructive and maintenance respiration in growing *Helianthus tuberosus* leaves. *Botanical Magazine (Tokyo)*, **91**, 43–56.

KLAPWIJK, D. (1979). Seasonal effects on the cropping-cycle of lettuce in glasshouses during the winter. *Scientia Horticulturae*, **11**, 371–7.

KOBAYASHI, S. (1975). Growth analysis of plant as an assemblage of internodal segments – a case of sunflower plants in pure stands. *Japanese Journal of Ecology*, **25**, 61–70.

KOLLER, H. R. (1971). Analysis of growth within distinct strata of the soybean community. *Crop Science*, **11**, 400–2.

KOLLER, H. R., NYQUIST, W. E. and CHORUSH, I. S. (1970). Growth analysis of the soybean community. *Crop Science*, **10**, 407–12.

KORNHER, A. (1971). Untersuchungen zur Stoffproduktion von Futterpflanzenbeständen. I. Wachtumsanalytische Untersuchungen an Beständen von Wiesenschwingle (*Festuca pratensis* Huds.) und Wiesenlieschgras (*Phleum pratense* L.) *Acta Agriculturae Scandinavica*, **21**, 215–36.

KOWALSKI, C. J. and GUIRE, K. E. (1974). Longitudinal data analysis. *Growth*, **38**, 131–69.

KREUSLER, U., PREHN, A. and BECKER, G. (1877a). Beobachtungen über das Wachsthum der Maispflanze. *Landwirtschaftliche Jahrbücher*, **6**, 759–86.

KREUSLER, U., PREHN, A. and BECKER, G. (1877b). Beobachtungen über das Wachsthum der Maispflanze (Bericht über die Versuche vom Jahre 1876). *Landwirtschaftliche Jahrbücher*, **6**, 787–800.

KREUSLER, U., PREHN, A. and HORNBERGER, R. (1878). Beobachtungen über das Wachsthum der Maispflanze (Bericht über die Versuche vom Jahre 1877). *Landwirtschaftliche Jahrbücher*, **7**, 536–64.

KREUSLER, U., PREHN, A. and HORNBERGER, R. (1879). Beobachtungen über das Wachsthum der Maispflanze (Bericht über die Versuche vom Jahre 1878). *Landwirtschaftliche Jahrbücher*, **8**, 617–22.

KVĚT, J. (1978). Growth analysis of fishpond littoral communities. In *Pond Littoral Systems: Structure and Functioning*. Ecological Studies, 28. Ed. D. DYKYJOVÁ and J. KVĚT, pp. 198–291. Springer Verlag, Berlin.

KVĚT, J. and ONDOK, J. P. (1971). The significance of biomass duration. *Photosynthetica*, **5**, 417–20.

KVĚT, J., ONDOK, J. P., NEČAS, J. and JARVIS, P. G. (1971). Methods of growth analysis. In *Plant Photosynthetic Production: Manual of Methods*. Ed. Z. ŠESTÁK, J. ČATSKÝ and P. G. JARVIS, pp. 343–91. Dr. W. Junk N.V., The Hague.

KVĚT, J., SVOBODA, J. and FIALA, K. (1969). Canopy development in stands of *Typha latifolia* L. and *Phragmites communis* Trin. in South Moravia. *Hidrobiologia*, **10**, 63–75.

LAMONT, B. B. and DOWNES, S. (1979). The longevity, flowering and fire history of the grasstrees *Xanthorrhoea preissii* and *Kingia australis*. *Journal of Applied Ecology*, **16**, 893–9.

LANDSBERG, J. J. (1974). Apple fruit bud development and growth analysis and an empirical model. *Annals of Botany*, **38**, 1013–23.

LANDSBERG, J. J. (1977). Some useful equations for biological studies. *Experimental Agriculture,* **13**, 273–86.

LAVAL-MARTIN, D. A., SHUCH, D. J. and EDMUNDS, L. N. Jr. (1979). Cell cycle-related and endogenously controlled circadian photosynthetic rhythms in *Euglena. Plant Physiology,* **63**, 495–502.

LAWN, R. J. (1979a). Agronomic studies on *Vigna* spp. in south-eastern Queensland. I. Phenological response of cultivars to sowing date. *Australian Journal of Agricultural Research,* **30**, 855–70.

LAWN, R. J. (1979b). Agronomic studies on *Vigna* spp. in south-eastern Queensland. II. Vegetative and reproductive response of cultivars to sowing date. *Australian Journal of Agricultural Research,* **30**, 871–82.

LEDIG, F. T. (1969). A growth model for tree seedlings based on the rate of photosynthesis and the distribution of photosynthate. *Photosynthetica,* **3**, 263–75.

LEDIG, F. T. and PERRY, T. O. (1969). Net assimilation rate and growth in loblolly pine seedlings. *Forestry Science,* **15**, 431–8.

LEOPOLD, A. C. and KRIEDMAN, P. E. (1975). *Plant Growth and Development,* second edition. McGraw Hill, New York.

LESHEM, Y., THAINE, R., HARRIS, C. E. and CANAWAY, R. J. (1972). Water loss from cut grass with special reference to hay-making. *Annals of Applied Biology,* **72**, 89–104.

LINHART, Y. B. and WHELAN, R. J. (1980). Woodland regeneration in relation to grazing and fencing in Coed Gorswen, North Wales. *Journal of Applied Ecology,* **17**, 827–40.

LIORET, C. (1974). L'analyse des courbes de croisssance. *Physiologie Végétale,* **12**, 413–34.

LITTLETON, E. J., DENNETT, M. D., ELSTON, J. and MONTEITH, J. L. (1979a). The growth and development of cowpeas (*Vigna unguiculata*) under tropical field conditions. 1. Leaf area. *Journal of Agricultural Science,* **93**, 291–307.

LITTLETON, E. J., DENNET, M. D., MONTEITH, J. L. and ELSTON, J. (1979b). The growth and development of cowpeas (*Vigna unguiculata*) under tropical field conditions. 2. Accumulation and partition of dry weight. *Journal of Agricultural Science,* **93**, 309–20.

LUPTON, F. G. H., ALI, M. A. M. and SUBRAMANIAM, S. (1967). Varietal differences in growth parameters of wheat and their importance in determining yield. *Journal of Agricultural Science,* **69**, 111–23.

MacARTHUR, R. H. and WILSON, E. D. (1967). *The Theory of Island Biogeography.* Princeton University Press, Princeton, N.J.

MacCOLL, D. (1977). Growth and sugar accumulation of sugarcane. II. Dry weight increments and estimates of assimilation rate. *Experimental Agriculture,* **13**, 61–9.

McCOLLUM, R. E. (1978). Analysis of potato growth under differing P regimes. II. Time by P-status interactions for growth and leaf efficiency. *Agronomy Journal,* **70**, 58–67.

MACDOWALL, F. D. H. (1972a). Growth kinetics of Marquis wheat. I. Light dependence. *Canadian Journal of Botany,* **50**, 89–99.

MACDOWALL, F. D. H. (1972b). Growth kinetics of Marquis wheat. II. Carbon dioxide dependence. *Canadian Journal of Botany,* **50**, 883–9.

MACDOWALL, F. D. H. (1972c). Growth kinetics of Marquis wheat. III. Nitrogen dependence. *Canadian Journal of Botany* **50**, 1749–61.

MACDOWALL, F. D. H. (1973a). Growth kinetics of Marquis wheat. IV. Temperature dependence. *Canadian Journal of Botany,* **51**, 729–36.

MACDOWALL, F. D. H. (1973b). Growth kinetics of Marquis wheat. V. Morphogenic dependence. *Canadian Journal of Botany,* **51**, 1259–65.

MACDOWALL, F. D. H. (1974). Growth kinetics of Marquis wheat. VI. Genetic dependence and winter hardening. *Canadian Journal of Botany,* **52**, 151–7.

McGREEVY, M. (1978). Incremental growth analysis of *Pseudotsuga menziesii. New Zealand Forestry Service, Forestry Research Institute, Symposium No. 15,* 155–70.

MACHIN, D. (1976). *Biomathematics: an Introduction.* Macmillan, London.

McKEE, G. W., LEE, H. J., KNIEVAL, D. P. and HOFFMAN, D. (1979). Rate of fill and length of the grain fill period for nine cultivars of spring oats. *Agronomy Journal,* **71**, 1029–34.

McKINNON, J. M., HESKETH, J. D. and BAKER, D. N. (1974). Analysis of the exponential growth equation. *Crop Science,* **14**, 549–51.

McKINNON, J. C. (1979). Energy allocation during growth of six maize hybrids in Nova Scotia. *Canadian Journal of Plant Science,* **59**, 667–77.

MAEDA, K. (1972). Growth analysis on the plant type in peanut varieties, *Arachis hypogaea* L. IV. Relationship between the varietal differences of the progress of leaf emergence on the main stem during pre-flowering period and the degree of morphological differentiation of leaf primordia in the embryo. *Proceedings of the Crop Science Society of Japan,* **41**, 179–86.

MAHMOUD, A., EL-SHEIKH, A. M., BASET, S. A. and HUNT, R. (1981). Temperature and the vegetative growth of two desert *Acacias. Annals of Botany,* **48**, 673–703.

MAJOR, D. J. (1980). Effect of simulated frost injury induced by paraquat on kernel growth and development in corn. *Canadian Journal of Plant Science,* **60**, 419–26.

MAKRIDAKIS, S. (1976). A survey of time series. *International Statistical Review,* **44**, 29–70.

MARANI, A. (1979). Growth rate of cotton bolls and their components. *Field Crops Research,* **2**, 169–75.

MARQUARDT, D. W. (1963). An algorithm for least squares estimation of non-linear parameters. *Journal of the Society of Industrial and Applied Mathematics,* **11**, 431–41.

MARRS, R. H., ROBERTS, R. D. and BRADSHAW, A. D. (1980). Ecosystem development on reclaimed china clay wastes. I. Assessment of vegetation and capture of nutrients. *Journal of Applied Ecology,* **17**, 709–17.

MASON, C. F. (1977). *Decomposition.* Studies in Biology No. 74. Edward Arnold, London.

MÁTHÉ, I. Jr. and MÁTHÉ, I. (1979). Comparative study on the alkaloid production of *Solanum dulcamara* chemotaxa during the vegetation period. *Acta Agronomica Academiae Scientiarum Hungaricae,* **28,** 458–66.

MATHER, K. (1964). *Statistical Analysis in Biology,* fifth edition. Methuen, London.

MATTHEWS, S. (1973). The effect of time of harvest on the viability and pre-eminence mortality in soil of pea (*Pisum sativum* L.) seeds. *Annals of Applied Biology,* **73,** 211–9.

MAUNY, J. R., FRY, K. E. and GUINN, G. (1978). Relationship of photosynthetic rate to growth and fruiting of cotton. *Crop Science,* **18,** 259–63.

MAX, T. A. and BURKHART, H. E. (1976). Segmented polynomial regression applied to taper regressions. *Forest Science,* **22,** 283–9.

MAYNARD SMITH, J. (1968). *Mathematical Ideas in Biology.* Cambridge University Press, London.

MEAD. R. (1971). A note on the use and misuse of regression models in ecology. *Journal of Ecology,* **59,** 215–9.

MEAD, R. and PIKE, D. J. (1975). A review of response surface methodology from a biometric viewpoint. *Biometrics,* **31,** 803–51.

MELSTED, S. W. and PECK, T. R. (1977). The Mitscherlich-Bray Growth Function. In *Soil Testing: Correlating and Interpreting the Analytical Results* (ASA Special Publication Number 29). Eds T. R. PECK, J. T. COPE and D. A. WHITNEY, pp. 1–18. American Society of Agronomy, Wisconsin.

MENGEL, D. B. and BARBER, S. A. (1974). Rate of nutrient uptake per unit of corn root under field conditions. *Agronomy Journal,* **66,** 399–402.

MEYER, J. L. (1980). Dynamics of phosphorus and organic matter during leaf decomposition in a forest stream. *Oikos,* **34,** 44–53.

MICHELINI, F. J. (1958). The plastochron index in developmental studies of *Xanthium italicum* Moretti. *American Journal of Botany,* **45,** 525–33.

MIDDLETON, W., JARVIS, B. C. and BOOTH, A. (1980). The role of leaves in auxin and boron-dependent rooting of stem cuttings of *Phaseolus aureus* Roxb. *New Phytologist,* **84,** 251–9.

MIGUS, W. N. and HUNT, L. A. (1980). Gas exchange rates and nitrogen concentrations in two winter wheat cultivars during the grain-filling period. *Canadian Journal of Botany,* **58,** 2110–16.

MILNER, C. and HUGHES, R. E. (1968). *Methods for the Measurement of the Primary Production of Grassland.* IBP Handbook No. 6. Blackwell Scientific Publications, Oxford.

MILTHORPE, F. L. and MOORBY, J. (1980). *An Introduction to Crop Physiology,* second edition. Cambridge University Press, London.

MISRA, R. K. (1980). Statistical comparisons of several growth curves of the von Bertalanffy type. *Canadian Journal of Fisheries and Aquatic Sciences,* **37,** 920–6.

MONSELISE, S. P., VARGA, A. and BRUINSMA, J. (1978). Growth analysis of the tomato fruit, *Lycopersicon esculentum* Mill. *Annals of Botany,* **42,** 1245–7.

MONTENEGRO, G., ALJARO, M. E. and KUMMEROW, J. (1979). Growth dynamics of Chilean matorral shrubs. *Botanical Gazette,* **140**, 114–9.

MONYO, J. H. and WHITTINGTON, W. J. (1970). Genetic analysis of root growth in wheat. *Journal of Agricultural Science,* **74**, 329–38.

MONYO, J. H. and WHITTINGTON, W. J. (1971). Inheritance of plant growth characters and their relation to yield in wheat substitution lines. *Journal of Agricultural Science,* **76**, 167–72.

MOORBY, J. (1970). The production, storage and translocation of carbohydrates in developing potato plants. *Annals of Botany,* **34**, 297–308.

MORAN, P. A. P. (1974). How to find out in statistical and probability theory. *International Statistical Review,* **42**, 299–303.

MORGAN, W. C. and PARBERY, D. G. (1977). Effects of *Pseudopeziza* leaf spot disease on growth and yield in lucerne. *Australian Journal of Agricultural Research,* **28**, 1029–40.

MORLEY, F. H. W. (1968). Pasture growth curves and grazing management. *Australian Journal of Experimental Agriculture and Animal Husbandry,* **8**, 40–5.

MUKAI, H., AIOI, K., KOIKE, I., IISUMI, H., OHTSU, H. and HATTORI, A. (1979). Growth and organic production of eelgrass (*Zostera marina* L.) in temperate waters of the Pacific coast of Japan. I. Growth analysis in spring–summer. *Aquatic Botany,* **7**, 47–56.

MURAMOTO, H., HESKETH, J. and EL-SHARKAWY, M. (1965). Relationships among rate of leaf area development, photosynthetic rate, and rate of dry matter production among American cultivated cottons and other species. *Crop Science,* **5**, 163–6.

MURATA, Y. (1975a). The effect of solar radiation, temperature, and aging on net assimilation rate of crop stands – from the analysis of the 'Maximal Growth Rate Experiment' of IBP/PP. I. The case of rice plants. *Proceedings of the Crop Science Society of Japan,* **44**, 153–9.

MURATA, Y. (1975b). The effect of solar radiation, temperature and aging on net assimilation rate of crop stands – from the analysis of the 'Maximal Growth Rate Experiment' of IBP/PP. II. The case of maize and soybean plants. *Proceedings of the Crop Science Society of Japan,* **44**, 160–5.

NAMKOONG, G. and MATZINGER, D. F. (1975). Selection for annual growth curves in *Nicotiana tabacum* L. *Genetics,* **81**, 377–86.

NÁTR, L. and KOUSALOVÁ, I. (1965). Comparison of results of photosynthetic intensity measurements in cereal leaves as determined by the dry weight increase or by the gazometric method. *Biologia Plantarum (Praha),* **7**, 98–108.

NÁTR, L., APEL, P. and KOUSALOVÁ, I. (1978). Mathematical description of nitrogen and phosphorus accumulation in developing barley kernels. *Biologia Plantarum (Praha),* **20**, 248–55.

NEALES, T. F. and NICHOLLS, A. O. (1978). Growth responses of young wheat plants to a range of ambient CO_2 levels. *Australian Journal of Plant Physiology,* **5**, 45–59.

NEČAS, J. (1974). Physiological approach to the analysis of some complex characters of potatoes. *Potato Research,* **17**, 3–23.

NEČAS, J., ZRŮST, J. and PARTYKOVÁ, E. (1967). Determination of the leaf area of potato plants. *Photosynthetica,* **1**, 97–111.

NELDER, J. A. (1961). The fitting of a generalization of the logistic curve. *Biometrics,* **17**, 89–110.

NELDER, J. A. (1962). An alternative form of a generalized logistic equation. *Biometrics,* **18**, 614–16.

NELDER, J. A. (1975). *Computers in Biology.* Wykeham, London.

NELDER, J. A., AUSTIN, R. B., BLEASDALE, J. K. A. and SALTER, P. J. (1960). An approach to the study of yearly and other variation in crop yields. *Journal of Horticultural Science,* **35**, 73–82.

NICHOLLS, A. O. and CALDER, D. M. (1973). Comments on the use of regression analysis for the study of plant growth. *New Phytologist,* **72**, 571–81;

NICHOLS, M. A. (1972). The effect of fertilizers on the growth of lettuce in New Zealand. *Horticultural Research,* **12**, 107–118.

NIE, N. H., HULL, C. H., JENKINS, J. G., STEINBRENNER, K. and BENT, D. H. (1975). *SPSS: Statistical Package for the Social Sciences,* second edition. McGraw-Hill, New York.

NYE, P. H., BREWSTER, J. L. and BHAT, K. K. S. (1975). The possibility of predicting solute uptake and plant growth response from independently measured soil and plant characteristics. *Plant and Soil,* **42**, 161–70.

OBRUCHEVA, N. V. and KOVALEV, A. G. (1979). Physiological interpretation of sigmoid growth curves of plant organs. *Fiziologiya Rastenii,* **26**, 1029–43.

OJEHOMON, O. O. (1970). A comparison of the vegetative growth, development and seed yield of three varieties of cowpea, *Vigna unguiculata* (L.) Walp. *Journal of Agricultural Science,* **74**, 363–74.

ONDOK, J. P. (1971a). Calculation of mean leaf area ratio in growth analysis. *Photosynthetica,* **5**, 269–71.

ONDOK, J. P. (1971b). Indirect estimation of primary values used in plant growth analysis. In *Plant Photosynthetic Production: Manual of Methods.* Eds Z. ŠESTÁK, J. ČATSKÝ and P. G. JARVIS, pp. 392–411. Dr. W. Junk N.V., The Hague.

ONDOK, J. P. (1978). Estimation of net photosynthetic efficiency from growth analytical data. In *Pond Littoral Systems: Structure and Functioning* (Ecological Studies, 28). Eds D. DYKYJOVÁ and J. KVĚT, pp. 221–91. Springer-Verlag, Berlin.

ONDOK, J. P. and KVĚT, J. (1971). Integral and differential formulae in growth analysis. *Photosynthetica,* **5**, 358–63.

ORMROD, D. P., HAMMER, P. A., KRIZEK, D. T., TIBBITTS, T. W., McFARLANE, J. C. and LANGHANS, R. W. (1980). Base-line growth studies of 'First Lady' Marigold in controlled environments. *Journal of the American Society for Horticultural Science,* **105**, 632–8.

PARKER, R. E. (1978). *Introductory Statistics for Biology,* second edition. Studies in Biology No. 43. Edward Arnold, London.

PARSONS, I. T. and HUNT, R. (1981). Plant growth analysis: a curve-fitting program using the method of *B*-splines. *Annals of Botany,* **48**, 341–52.

PASSIOURA, J. B. (1979). Accountability, philosophy and plant physiology. *Search*, **10**, 347–50.

PATTERSON, R. P., RAPER, C. F. Jr. and GROSS, H. D. (1979). Growth and specific nodule activity of soybean during application and recovery of a leaf moisture stress. *Plant Physiology*, **64**, 551–6.

PAYANDEH, B., WALLACE, D. R. and MacLEOD, D. M. (1980). An empirical regression function suitable for modelling spore germination subject to temperature threshold. *Canadian Journal of Botany*, **58**, 936–41.

PEARL, R. and REED, L. J. (1923). On the mathematical theory of population growth. *Metron*, **3**, 1–15.

PEARL, R., WINSOR, A. A. and MINER, J. R. (1928). The growth of seedlings of the canteloup, *Cucumis melo*, in the absence of exogenous food and light. *Proceedings of the National Academy of Sciences*, **14**, 1–4.

PEARSON, R. C., ALDWINCKLE, H. S. and SEEM, R. C. (1977). Teliospore germination and basidiospore formation in *Gymnosporangium juniperi-virginianae:* a regression model of temperature and time effects. *Canadian Journal of Botany*, **55**, 2832–7.

PEGELOW, E. J., TAYLOR, B. B., HORROCKS, R. D., BUXTON, D. R., MARX, D. B. and WANJURA, D. F. (1977). The Gompertz function as a model for cotton hypocotyl elongation. *Agronomy Journal*, **69**, 875–8.

PETERSEN, R. G. (1977). Use and misuse of multiple comparison procedures. *Agronomy Journal*, **69**, 205–8.

PHATAK, S. C., GLAZE, N. C., DOWLER, C. C. and THREADGILL, E. D. (1980). Responses of turnip greens to preplant tillage treatment. *Journal of the American Society of Horticultural Science*, **105**, 556–9.

PHILLIPSON, J. (1966). *Ecological Energetics*. Studies in Biology No. 1. Edward Arnold, London.

PIANKA, E. R. (1970). On *r*- and *K*-selection. *American Naturalist*, **104**, 592–7.

PICARD, D., COUCHAT, P. and MOUTONNET, P. (1980). Données préliminaires sur la transpiration du riz pluvial, variété IRAT 13, soumis à une carence hydrique. *Plant and Soil*, **57**, 423–30.

PIENAAR, L. V. and TURNBULL, K. J. (1973). The Chapman-Richards generalization of von Bertalanffy's growth model for basal area growth and yield in even-aged stands. *Forest Science*, **19**, 2–22.

POLLARD, J. H. (1977). *A Handbook of Numerical and Statistical Techniques*. Cambridge University Press, London.

PORTERFIELD, W. M. (1928). A study of the grand period of growth in bamboo. *Bulletin of the Torrey Botanical Club*, **55**, 327–405.

PORTSMOUTH, G. C. (1937). The effect of alternate periods of light and darkness of short duration on the growth of the cucumber. *Annals of Botany*, **1**, 175–89.

POTTER, J. R. and JONES, J. W. (1977). Leaf area partitioning as an important factor in growth. *Plant Physiology*, **59**, 10–14.

PRÉCSÉNYI, I., CZIMBER, G., CSALA, G., SZŐCS, Z., MOLNÁR, E. and MELKÓ, E. (1976). Studies on the growth analysis of maize hybrids (OSSK-218 and DKXL-342). *Acta Botanica Academiae Scientarum Hungaricae*, **22**, 185–200.

PREECE, M. A. and BAINES, M. J. (1978). A new family of mathematical models describing the human growth curve. *Annals of Human Biology,* **5**, 1–24.

PRESCOTT, J. A. (1922). The flowering curve of the Egyptian cotton plant. *Annals of Botany,* **36**, 121–30.

PROMNITZ, L. C. (1975). A photosynthetic allocation model for tree growth. *Photosynthetica,* **9**, 1–15.

PRUITT, K. M., DeMUTH, R. E. and TURNER, M. E. Jr. (1979). Practical application of the generic growth theory and the significance of the growth curve parameters. *Growth,* **43**, 19–35.

PULLI, S. (1980a). Development and productivity of timothy (*Phleum pratense* L.). *Journal of the Scientific Agricultural Society of Finland,* **52**, 368–92.

PULLI, S. (1980b). Growth factors and management technique used in relation to the developmental rhythm and yield formation pattern of a pure grass stand. *Journal of the Scientific Agricultural Society of Finland,* **52**, 281–330.

PULLI, S. (1980c). Growth factors and management technique used in relation to the developmental rhythm and yield formation pattern of a clover-grass stand. *Journal of the Scientific Agricultural Society of Finland,* **52**, 191–214.

RADFORD, P. J. (1967). Growth analysis formulae – their use and abuse. *Crop Science,* **7**, 171–5.

RAJA HARUN, R. M. and BEAN, E. W. (1979). Seed development and seed shedding in North Italian ecotypes of *Lolium multiflorum. Grass and Forage Science,* **34**, 215–20.

RAO, N. G. and MURTY, B. R. (1963). Growth analysis in grain sorghums of Deccan. *Indian Journal of Agricultural Science,* **33**, 155–62.

RAPER, C. D. Jr. (1977). Relative growth and nutrient accumulation rates for tobacco. *Plant and Soil,* **46**, 473–86.

RAPER, C. D. Jr., WEEKS, W. W. and WANN, M. (1976). Temperatures in early post-transplant growth: influence on carbohydrate and nitrogen utilization and distribution in tobacco. *Crop Science,* **16**, 753–57.

RAPER, C. D. Jr., PARSONS, L. R., PATTERSON, D. T. and KRAMER, P. J. (1977). Relationship between growth and nitrogen accumulation for vegetative cotton and soybean plants. *Botanical Gazette,* **138**, 129–37.

RAWSON, H. M., CONSTABLE, G. A. and HOWE, G. N. (1980). Carbon production of sunflower cultivars in field and controlled environments. II. Leaf growth. *Australian Journal of Plant Physiology,* **7**, 575–86.

REES, A. R. (1963). An analysis of growth of oil palms under nursery conditions. II. The effect of spacing and season on growth. *Annals of Botany* **27**, 615–26.

REES, A. R. and CHAPAS, L. C. (1963). An analysis of growth of oil palms under nursery conditions. I. Establishment and growth in the wet season. *Annals of Botany,* **27**, 607–14.

REPKA, J. and KOSTREJ, A. (1968). Making use of the growth analysis in the evaluation of the agricultural experiment (in Czech). *Polnohospodárstvo,* **14**, 236–47.

RICHARDS, F. J. (1959). A flexible growth function for empirical use. *Journal of Experimental Botany,* **10**, 290–300.

RICHARDS, F. J. (1969). The quantitative analysis of growth. In *Plant Physiology – a Treatise.* VA. *Analysis of Growth: Behavior of Plants and their Organs.* Ed. F. C. STEWARD, pp. 1–76. Academic Press, London.

RICHARDS, O. W. (1928). The growth of the yeast *Saccharomyces cerevisiae.* I. The growth curve, its mathematical analysis, and the effect of temperature on the yeast growth. *Annals of Botany,* **42**, 271–83.

RIFFENBURGH, R. H. (1966). On growth parameter estimation for early life stages. *Biometrics,* **22**, 162–78.

ROACH, B. M. B. (1926). On the relation of certain soil algae to some soluble carbon compounds. *Annals of Botany,* **40**, 149–201.

ROACH, B. M. B. (1928). On the influence of light and of glucose on the growth of a soil alga. *Annals of Botany,* **42**, 317–45.

ROBSON, M. J. and PARSONS, A. J. (1978). Nitrogen deficiency in small closed communities of S24 ryegrass. I. Photosynthesis, respiration, dry matter production and partition. *Annals of Botany,* **42**, 1185–97.

RORISON, I. H. and GUPTA, P. L. (1974). The growth of seedlings in response to variable phosphorus supply. In *Plant Analysis and Fertilizer Problems* (Proceedings of the 7th International Colloquium, Hanover, Federal Republic of Germany, September 1974). Ed. J. WEHRMANN, pp. 373–82. German Society of Plant Nutrition.

RORISON, I. H., PETERKIN, J. H. and CLARKSON, D. T. (1982). Nitrogen source, temperature and plant growth. In *Nitrogen as an Ecological Factor* (Symposium of the British Ecological Society, No. 22). Eds J. A. LEE and I. H. RORISON. Blackwell Scientific Publications, Oxford.

ROSE, D. A. and CHARLES-EDWARDS, D. A. (Eds) (1981). *Mathematics and Plant Physiology.* Academic Press, London.

RUDOLFS, R. (1927). Influence of salt upon growth rate of *Asparagus. Botanical Gazette,* **83**, 94–8.

RÜEGG, J. J. and ALSTON, A. M. (1978). Seasonal and diurnal variation of nitrogenase activity (acetylene reduction) in barrel medic (*Medicago truncatula* Gaertn.) grown in pots. *Australian Journal of Agricultural Research,* **29**, 951–62.

RUFTY, R. W., MINER, G. S. and RAPER, C. D. Jr. (1979). Temperature effects on growth and manganese tolerance in tobacco. *Agronomy Journal,* **71**, 638–44.

RUMBERG, C. B., LUDWICK, A. E. and SIEMER, E. G. (1980). Increased flowering and yield from a bromegrass-timothy meadow by timing of nitrogen. *Agronomy Journal,* **72**, 103–7.

RYLE G. J. A., COBBY, J. M. and POWELL, C. E. (1976). Synthetic and maintenance respiratory losses of $^{14}CO_2$ in uniculm barley and maize. *Annals of Botany,* **40**, 571–86.

SAKI, T. (1965). Growth analysis of plants (in Japanese). *Botanical Magazine (Tokyo),* **78**, 111–9.

SANDERS, F. E., TINKER, P. B., BLACK, R. L. B. and PALMERLEY, S. M. (1977). The development of endomycorrhizal root systems: I. Spread of infection and growth-promoting effects with four species of vesicular-arbuscular endophyte. *New Phytologist,* **78**, 257–68.

SANDLAND, R. L. and MCGILCHRIST, C. A. (1979). Stochastic growth curve analysis. *Biometrics,* **35**, 255–71.

SARUKHÁN, J. and HARPER, J. L. (1973). Studies on plant demography: *Ranunculus repens* L., *R. bulbosus* L. and *R. acris.* I. Population flux and survivorship. *Journal of Ecology,* **61**, 675–716.

SAVINOV, I. P., VASILYEV, B. R. and SCHMIDT, V. M. (1977). On one class of the plant growth curves. *Zhurnal Obschcheib Biologii,* **38**, 659–64.

SCHOENBERG, I. J. (1946). Contributions to the problem of approximation of equidistant data by analytic functions. Part A. On the problem of smoothing or graduation. A first class of analytic approximation formulae. *Quarterly of Applied Mathematics,* **4**, 45–99.

SCOTT, D. (1970). Relative growth rates under controlled temperatures of some New Zealand indigenous and introduced grasses. *New Zealand Journal of Botany,* **8**, 76–81.

SCOTT, R. K., ENGLISH, S. D., WOOD, D. W. and UNSWORTH, M. H. (1973). The yield of sugar beet in relation to weather and length of growing season. *Journal of Agricultural Science,* **81**, 339–47.

SCOTTER, D. R., CLOTHIER, B. E. and TURNER, M. A. (1979). The soil water balance in a Fragiaqualf and its effect on pasture growth in Central New Zealand. *Australian Journal of Soil Research,* **17**, 455–65.

SEMU, E. and HUME, D. J. (1979). Effects of inoculation and fertilizer N levels on N_2 fixation and yields of soybeans in Ontario. *Canadian Journal of Plant Science,* **59**, 1129–37.

ŠESTÁK, Z., ČATSKÝ, J. and JARVIS, P. G. (Eds) (1971). *Plant Photosynthetic Production: Manual of Methods.* Dr. W. Junk N. V., The Hague.

ŠESTÁK, Z., JARVIS, P. G. and ČATSKÝ, J. (1971). Criteria for the selection of suitable methods. In *Plant Photosynthetic Production: Manual of Methods.* Eds Z. ŠESTÁK, J. ČATSKÝ and P. G. JARVIS, pp. 1–48. Dr. W. Junk N. V., The Hague.

ŠETLÍK, I. and ŠESTÁK, Z. (1971). Use of leaf tissue samples in ventilated chambers for long-term measurements of photosynthesis. In *Plant Photosynthetic Production: Manual of Methods.* Eds Z. ŠESTÁK, J. ČATSKÝ and P. G. JARVIS, pp. 316–42. Dr. W. Junk N. V., The Hague.

SILSBURY, J. H. (1969). Seedling growth of summer-dormant and non-dormant ryegrass in relation to temperature. *Australian Journal of Agricultural Research,* **20**, 417–23.

SILSBURY, J. H. (1971). The effects of temperature and light energy on dry weight and leaf area changes in seedling plants of *Lolium perenne* L. *Australian Journal of Agricultural Research,* **22**, 177–87.

SILSBURY, J. H. (1979). Growth, maintenance and nitrogen fixation of nodulated plants of subterranean clover (*Trifolium subterraneum* L.). *Australian Journal of Plant Physiology,* **6**, 165–76.

SILSBURY, J. H., ADEM, L., BAGHURST, P. and CARTER, E. D. (1979). A quantitative examination of the growth of swards of *Medicago truncatula* cv. Jemalong. *Australian Journal of Agricultural Research,* **30**, 53–63.

SILSBURY, J. H. and FUKAI, S. (1977). Effects of sowing time and sowing density on the growth of subterranean clover at Adelaide. *Australian Journal of Agricultural Research,* **28**, 427–40.

SIMMONS, S. R. and CROOKSTON, R. K. (1979). Rate and duration of growth of kernels formed at specific florets in spikelets of spring wheat. *Crop Science,* **19**, 690–3.

SIMPSON, L. A. and GUMBS, F. A. (1980). Effect of soil micro-relief of a Lousiana bank system of field drainage after conversion from the cambered bed system on sugar cane. *Agronomy Journal,* **72**, 465–9.

SIMS, R. E. H. (1979a). Comparative methods of harvesting oilseed rape. *New Zealand Journal of Experimental Agriculture,* **7**, 79–83.

SIMS, R. E. H. (1979b). Drying cycles and optimum harvest stage of oilseed rape. *New Zealand Journal of Experimental Agriculture,* **7**, 85–89.

SINGH, D., SINGH, H. P., SINGH, P. and JHA, M. P. (1979). A study of pre-harvest forecasting of yield of jute. *Indian Journal of Agricultural Research,* **13**, 167–79.

SINGH, J. S., LAUENROTH, W. K. and STEINHORST, R. K. (1975). Review and assessment of various techniques for estimating net aerial primary production in grasslands from harvest data. *Botanical Review,* **41**, 181–232.

SIVAKUMAR, M. V. K. (1978). Prediction of leaf area index in Soya bean (*Glycine max* (L.) Merrill). *Annals of Botany,* **42**, 251–3.

SIVAKUMAR, M. V. K. and SHAW, R. H. (1978). Methods of growth analysis in field-grown soya beans (*Glycine max* (L.) Merrill). *Annals of Botany,* **42**, 213–22.

SMITH, D. and KEYFITZ, N. (1977). *Mathematical Demography: Selected Papers.* Biomathematics: Volume 6. Springer-Verlag, Berlin.

SMITH, D. L. and ROGAN, P. G. (1980). Correlative inhibition in the shoot of *Agropyron repens* (L.) Beauv. *Annals of Botany,* **46**, 285–96.

SMITH, G. S., MIDDLETON, K. R. and EDMONDS, A. S. (1978). A classification of pasture and fodder plants according to their ability to translocate sodium from their roots into aerial parts. *New Zealand Journal of Experimental Agriculture,* **6**, 183–6.

SMITH, J. M. (1977). *Scientific Analysis on the Pocket Calculator,* second edition. John Wiley, New York.

SMITH, P. R. and NEALES, T. F. (1977). The growth of young peach trees following infection by the viruses of peach rosette and decline disease. *Australian Journal of Agricultural Research,* **28**, 441–4.

SMYTH, D. A. and DUGGER, W. M. (1980). Effects of boron deficiency on ^{86}Rubidium uptake and photosynthesis in the diatom *Cylindeotheca fusiformis. Plant Physiology,* **66**, 692–5.

SOBULO, R. A. (1972). Studies on white yam (*Dioscorea rotundata*). I. Growth analysis. *Experimental Agriculture,* **8**, 99–106.

SOFIELD, I., EVANS, L. T., COOK, M. G. and WARDLAW, I. F. (1977). Factors influencing the rate and duration of grain filling in wheat. *Australian*

Journal of Plant Physiology, **4**, 785–91.

SOFIELD, I., WARDLAW, I. F., EVANS, L. T. and ZEE, S. Y. (1977). Nitrogen, phosphorus and water contents during grain development and maturation in wheat. *Australian Journal of Plant Physiology*, **4**, 799–810.

SOLÁROVÁ, J. (1980). Diffusive conductances of adaxial (upper) and abaxial (lower) epidermes: response to quantum irradiance during development of primary *Phaseolus vulgaris* L. leaves. *Photosynthetica*, **14**, 523–31.

SOLOMON, M. E. (1976). *Population Dynamics*, second edition. Studies in Biology No. 18. Edward Arnold, London.

SORENSEN, F. C. (1978). Date of sowing and nursery growth of provenances of *Pseudotsuga menziensii* given two fertilizer regimes. *Journal of Applied Ecology*, **15**, 273–9.

STANHILL, G. (1977). Allometric growth studies of the carrot crop. I. Effects of plant development and cultivar. *Annals of Botany*, **41**, 533–40.

STEER, B. T. and BLACKWOOD, G. C. (1978). Within-day changes in the polyribosome content and in synthesis of proteins in leaves of *Capsicum annuum* L. *Plant physiology*, **62**, 907–11.

STOY, V. (1965). Photosynthesis, respiration and carbohydrate accumulation in spring wheat in relation to yield. *Physiologia Plantarum*, **18** (*Supplementum IV*), 1–125.

STRIBLEY, D. P. and READ, D. J. (1975). Some nutritional aspects of the biology of ericaceous mycorrhizas. In *Endomycorrhizas*. Eds F. E. SANDERS, M. MOSSE and P. B. TINKER, pp. 195–207. Academic Press, London.

STRIBLEY, D. P. and READ, D. J. (1976). The biology of mycorrhiza in the Ericaceae. VI. The effects of mycorrhizal infection and concentration of ammonium nitrogen on growth of cranberry (*Vaccinium macrocarpon* Ait.) in sand culture. *New Phytologist*, **77**, 63–72.

STRIBLEY, D. P., READ, D. J. and HUNT, R. (1975). The biology of mycorrhiza in the Ericaceae. V. The effects of mycorrhizal infection, soil type and partial soil-sterilization (by gamma-irradiation) on growth of cranberry (*Vaccinium macrocarpon* Ait.). *New Phytologist*, **75**, 119–30.

STROH, J. R. (1971). Variation in growth curve increments. *Agronomy Journal*, **63**, 512–3.

SVÁB, J., MÁNDY, G. and BÓCSA, I. (1968). Method of growth analysis by logistic function in hemp. *Növénytermelés*, **17**, 285–92.

SZEICZ, G., van BAVEL, C. H. M. and TAKAMI, S. (1973). Stomatal factor in the water use and dry matter production by sorghum. *Agricultural Meteorology*, **12**, 361–89.

TAYLOR, G. B. (1972). The effect of seed size on seedling growth in subterranean clover (*Trifolium subterraneum* L.). *Australian Journal of Agricultural Research*, **23**, 595–603.

TERMAN, G. L. and NELSON, L. A. (1976). Comments on use of multiple regression in plant analysis interpretation. *Agronomy Journal*, **68**, 148–50.

THOMAS, A. W., SNYDER, W. M. and BRUCE, R. R. (1977). Smoothing, interpolation, and gradients from limited data. *Agronomy Journal*, **69**, 747–50.

THOMAS, H. and NORRIS, I. B. (1977). The growth responses of *Lolium perenne* to the weather during winter and spring at various altitudes in mid-Wales. *Journal of Applied Ecology,* **14**, 949–64.

THORNE, G. N. (1960). Variations with age in net assimilation rate and other growth attributes of sugar-beet, potato, and barley in a controlled environment. *Annals of Botany,* **24**, 356–71.

THORNE, G. N. (1961). Effects of age and environment on net assimilation rate of barley. *Annals of Botany,* **25**, 29–38.

THORNLEY, J. H. M. (1976). *Mathematical Models in Plant Physiology: a Quantitative Approach to Problems in Plant and Crop Physiology.* Academic Press, London.

THORNLEY, J. H. M. (1980). Research strategy in the plant sciences. *Plant, Cell and Environment,* **3**, 233–6.

THORNLEY, J. H. M. and HESKETH, J. D. (1972). Growth and respiration in cotton bolls. *Journal of Applied Ecology,* **9**, 315–7.

THORNLEY, J. H. M. and HURD, R. G. (1974). An analysis of the growth of young tomato plants in water culture at different light integrals and CO_2 concentrations. II. A mathematical model. *Annals of Botany,* **38**, 389–400.

TINKER, P. B. (1969). The transport of ions in the soil around plant roots. In *Ecological Aspects of the Mineral Nutrition of Plants.* Ed. I. H. RORISON, pp. 135–47. Blackwell Scientific Publications, Oxford.

TOTTMAN, D. R., MAKEPEACE, R. J. and BROAD, H. (1979). An explanation of the decimal code for the growth stages of cereals, with illustrations. *Annals of Applied Biology,* **93**, 221–34.

TOWNSEND, L. R. and MCRAE, K. B. (1980). The effect of nitrification inhibitor nitrapyrin on yield and on nitrogen fractions in soil and tissue of corn (*Zea mays* L.) grown in the Annapolis Valley of Nova Scotia. *Canadian Journal of Plant Science,* **60**, 337–47.

TROUGHTON, A. (1965). Intra-varietal variation in *Lolium perenne. Euphytica,* **14**, 59–66.

TROUGHTON, A. (1967). The effect of mineral nutrition on the distribution of growth in *Lolium perenne. Annals of Botany,* **31**, 447–54.

TROUGHTON, A. (1968). Influence of genotype and mineral nutrition on the distribution of growth within plants of *Lolium perenne* L. grown in soil. *Annals of Botany,* **32**, 411–23.

TROUGHTON, A. (1971). The relationship between the relative growth rates of the shoot system, number of tillers and mean tiller size in *Lolium perenne. Annals of Applied Biology,* **68**, 193–202.

TROUGHTON, A. (1977). The rate of growth and partitioning of assimilates in young grass plants: a mathematical model. *Annals of Botany,* **41**, 553–65.

TULLBERG, J. N. and ANGUS, D. E. (1978). The effect of potassium carbonate solution on the drying of lucerne. I. Laboratory studies. *Journal of Agricultural Science,* **91**, 551–6.

VAN DE DIJK, S. J. (1980). Two ecologically distinct subspecies of *Hypochaeris radicata* L. II. Growth response to nitrate and ammonium, growth strategy and formative aspects. *Plant and Soil,* **57**, 111–22.

VARGA, A. and BRUINSMA, J. (1976). Roles of seeds and auxins in tomato fruit growth. *Zeitschrift für Pflanzenphysiologie,* **80**, 95–104.

VARTHA, E. W. (1973). Effects of defoliation and nutrients on growth *Poa trivialis* L. (*sic*) with perennial ryegrass. *New Zealand Journal of Agricultural Research,* **16**, 43–8.

VENUS, J. C. and CAUSTON, D. R. (1979a). Plant growth analysis: a re-examination of the methods of calculation of relative growth and net assimilation rates without using fitted functions. *Annals of Botany,* **43**, 633–8.

VENUS, J. C. and CAUSTON, D. R. (1979b). Plant growth analysis: the use of the Richards function as an alternative to polynomial exponentials. *Annals of Botany,* **43**, 623–32.

VENUS, J. C. and CAUSTON, D. R. (1979c). Confidence limits for Richards functions. *Journal of Applied Ecology,* **16**, 939–49.

VERNON, A. J. and ALLISON, J. C. S. (1963). A method of calculating net assimilation rate. *Nature,* **200**, 814 only.

VIL'YAMS, M. V., TSVETKOVA, I. V., DERENDYAEVA, T. A., IVANOVA, I. E. and MAKSIMOVA, É. V. (1979). Cultivation of beet plants under conditions of balanced mineral nutrition. *Soviet Plant Physiology,* **26**, 94–9.

VIRAGH, K. (1979). Wachstumsanalyse der Sonnen- und Schattenblätter von *Quercus cerris* und *Quercus petraea* (1973–1975). *Acta Botanica Academiae Scientiarum Hungaricae,* **25**, 143–64.

VOLDENG, H. D. and BLACKMAN, G. E. (1973). An analysis of the components of growth which determine the course of development under field conditions of selected inbreds and their hybrids of *Zea mays. Annals of Botany,* **37**, 539–52.

VYVYAN, M. C. (1924). Studies of the rate of growth of leaves by a photographic method. I. The determination of the rate of growth of first leaves of *Phaseolus vulgaris. Annals of Botany,* **38**, 59–103.

WADDINGTON, C. H. (1956). *Principles of Embryology.* George Allen & Unwin, London.

WAHBA, G. and WOLD, S. (1975). Periodic splines for spectral density estimation: the use of cross validation for determining the degree of smoothing. *Communications in Statistics,* **4**, 125–41.

WALLACH, D. (1975). The effect of environmental factors on the growth of a natural pasture. *Agricultural Meteorology,* **15**, 231–44.

WALLACH, D. (1978). A simple model of cotton yield development. *Field Crops Research,* **1**, 269–81.

WALLACH, D. (1980). An empirical mathematical model of a cotton crop subjected to damage. *Field Crops Research,* **3**, 7–25.

WALLACH, D. and GUTMAN, M. (1976). Environment-dependent logistic equations applied to natural pasture growth curves. *Agricultural Meteorology,* **16**, 389–404.

WALLACH, D., MARANI, A. and KLETTER, E. (1978). The relation of cotton crop growth and development to final yield. *Field Crops Research,* **1**, 283–94.

WALLÉN, B. (1980). Structure and dynamics of *Calluna vulgaris* on sand dunes in South Sweden. *Oikos,* **35**, 20–30.

WALTON, D. W. H. (1976). Dry matter production in *Acaena* (Rosaceae) on a subantarctic island. *Journal of Ecology*, **64**, 399–415.

WALTON, D. W. H. and SMITH, R. I. L. (1976). Some limitations on plant growth and development in tundra regions – an investigation using phytometers. *New Phytologist*, **76**, 501–10.

WARDLAW, I. F., SOFIELD, I. and CARTWRIGHT, P. M. (1980). Factors limiting the rate of dry matter accumulation in the grain of wheat grown at high temperature. *Australian Journal of Plant Physiology*, **7**, 387–400.

WARREN WILSON, J. (1981). Integrated analysis of growth, photosynthesis and light interception for single plants and stands. *Annals of Botany*, **48**, 507–512.

WATANABE, K. and TAKAHASHI, Y. (1979). Effects of fertilization level on the regrowth of orchardgrass. II. Growth analysis in process of regrowth. *Journal of Japanese Grassland Science*, **25**, 203–9.

WATSON, D. J. (1947). Comparative physiological studies on the growth of field crops. I. Variation in net assimilation rate and leaf area between species and varieties, and within and between years. *Annals of Botany*, **11**, 41–76.

WATSON, D. J. (1952). The physiological basis of variation in yield. *Advances in Agronomy*, **4**, 101–45.

WATSON, D. J. (1958). The dependence of net assimilation rate on leaf area index. *Annals of Botany*, **22**, 37–54.

WATSON, D. J. (1968). A prospect of crop physiology. *Annals of Applied Biology*, **62**, 1–9.

WATSON, D. J. (1971). Size, structure, and activity of the productive system of crops. In *Potential Crop Production*. Eds P. F. WAREING and J. P. COOPER, pp. 76–88. Heinemann, London.

WEBB, W. L. (1975). Dynamics of photoassimilated carbon in Douglas fir seedlings. *Plant Physiology*, **56**, 455–9.

WELBANK, P. J. (1962). The effects of competition with *Agropyron repens* and nitrogen- and water-supply on the nitrogen content of *Impatiens parviflora*. *Annals of Botany*, **26**, 361–373.

WELBANK, P. J. (1964). Competition for nitrogen and potassium in *Agropyron repens*. *Annals of Botany*, **28**, 1–16.

WEST, C., BRIGGS, G. E. and KIDD, F. (1920). Methods and significant relations in the quantitative analysis of plant growth. *New Phytologist*, **19**, 200–7.

WHITEHEAD, F. H. and MYERSCOUGH, P. J. (1962). Growth analysis of plants. The ratio of mean relative growth rate to mean relative rate of leaf area increase. *New Phytologist*, **61**, 314–21.

WIEGAND, C. L., RICHARDSON, A. J. and KANEMASU, E. T. (1979). Leaf area index estimates for wheat from LANDSAT and their implications for evapotranspiration and crop modelling. *Agronomy Journal*, **71**, 336–42.

WIGGLESWORTH, V. B. (1967). The religion of science. *Annals of Applied Biology*, **60**, 1–10.

WILD, A., WOODHOUSE, P. M. and HOPPER, M. J. (1979). A comparison between the uptake of potassium by plants from solutions of constant potassium concentration and during depletion. *Journal of Experimental Botany,* **30**, 697–704.

WILLIAMS, R. F. (1946). The physiology of plant growth with special reference to the concept of net assimilation rate. *Annals of Botany,* **10**, 41–72.

WILLIAMS, R. F. (1948). The effects of phosphorus supply on the rates of intake of phosphorus and nitrogen and upon certain aspects of phosphorus metabolism in gramineous plants. *Australian Journal of Scientific Research, Series B,* **1**, 333–61.

WILLIAMS, R. F. (1964). The quantitative description of growth. In *Grasses and Grasslands.* Ed. C. BARNARD, pp. 89–101. Macmillan, London.

WILLIAMS, R. F. (1975). *The Shoot Apex and Leaf Growth.* Cambridge University Press, London.

WILLIAMS, R. F. and BOUMA, D. (1970). The physiology of growth in subterranean clover. I. Seedling growth and the pattern of growth at the shoot apex. *Australian Journal of Botany,* **18**, 127–48.

WILLMS, W., McLEAN, A. and KALNIN, C. (1980). Nutritive characteristics of grasses on spring range in South Central British Columbia in relation to time, habitat and fall grazing. *Canadian Journal of Plant Science,* **60**, 131–7.

WILSON, J. H., CLOWES, M. St.J. and ALLISON, J. C. S. (1973). Growth and yield of maize at different altitudes in Rhodesia. *Annals of Applied Biology,* **73**, 77–84.

WINSOR, C. P. (1932). The Gompertz curve as a growth curve. *Proceedings of the National Academy of Sciences of the United States of America,* **18**, 1–8.

WOLD, S. (1974). Spline functions in data analysis. *Technometrics,* **16**, 1–11.

WOODWARD, F. I. (1980). Review of HUNT, R. (1978): *Plant Growth Analysis. Journal of Applied Ecology,* **17**, 516 only.

WOODWARD, R. G. (1976). Photosynthesis and expansion of leaves of soybean grown in two environments. *Photosynthetica,* **10**, 274–9.

WOOLHOUSE, H. W. (1980). Review of HUNT, R. (1978): *Plant Growth Analysis. Journal of Ecology,* **68**, 320–1.

YANG, R. C., KOZAK, A. and SMITH, J. H. G. (1978). The potential of Weibull-type functions as flexible growth curves. *Canadian Journal of Forest Research,* **8**, 424–31.

YARRANTON, G. A. (1969). Plant ecology: a unifying model. *Journal of Ecology,* **57**, 245–50.

YOUNG, J. E. (1981). The use of canonical correlation analysis in the investigation of relationships between plant growth and environmental factors. *Annals of Botany,* **48**, 811–25.

YOUNG, J. K., WHISLER, F. D. and HODGES, H. F. (1979). Soybean leaf N as influenced by seedbed preparation methods and stages of growth. *Agronomy Journal,* **71**, 568–73.

YUKIMURA, T. and KANDA, M. (1978). Studies on population establishment in artificial grasslands. (a) Significance of seed weight as a factor influencing the neighbouring interaction during the period of grass establishment. *Reports of the Institute for Agricultural Research, Tohoku University,* **29**, 1–8.

ZADOKS, J. C., CHANG, T. T. and KOZAK, C. F. (1974). A decimal code for the growth stages of cereals. *Weed Research,* **14**, 415–21.

ZAR, J. H. (1967). The effect of changes in units of measurement on least squares regression lines. *BioScience,* **17**, 818–9.

ZAR, J. H. (1968). Calculation and miscalculation of the allometric equation as a model in biological data. *BioScience,* **18**, 1118–20.

ŻELAWSKI, W. and LECH, A. (1979). Growth function characterizing dry matter accumulation of plants. *Bulletin de l'Académie Polonaise des Sciences, Série des Sciences Biologiques* V, **27**, 675–81.

ŻELAWSKI, W. and LECH, A. (1980). Logistic growth function and their applicability for characterizing dry matter accumulation in plants. *Acta Physiologiae Plantarum,* **2**, 187–94.

ZRŮST, J., PARTYKOVÁ, E. and NEČAS, J. (1974). Relationships of leaf area to leaf weight and length in potato plants. *Phytosynthetica,* **8**, 118–24.

ZUCKERMANN, S. (Ed.) (1950). A discussion on the measurement of growth and form. *Proceedings of the Royal Society,* **B137**, 433–523.

Author index

Bold numbers refer to pages containing an illustration in the form of a quotation, equation, table entry, or text figure.

Systematic index

Subject index

Bold numbers indicate pages on which a main section begins; this may extend over more than one page.